Springer Theses

Recognizing Outstanding Ph.D. Research

For further volumes:
http://www.springer.com/series/8790

Aims and Scope

The series "Springer Theses" brings together a selection of the very best Ph.D. theses from around the world and across the physical sciences. Nominated and endorsed by two recognized specialists, each published volume has been selected for its scientific excellence and the high impact of its contents for the pertinent field of research. For greater accessibility to non-specialists, the published versions include an extended introduction, as well as a foreword by the student's supervisor explaining the special relevance of the work for the field. As a whole, the series will provide a valuable resource both for newcomers to the research fields described, and for other scientists seeking detailed background information on special questions. Finally, it provides an accredited documentation of the valuable contributions made by today's younger generation of scientists.

Theses are accepted into the series by invited nomination only and must fulfill all of the following criteria

- They must be written in good English.
- The topic should fall within the confines of Chemistry, Physics, Earth Sciences, Engineering and related interdisciplinary fields such as Materials, Nanoscience, Chemical Engineering, Complex Systems and Biophysics.
- The work reported in the thesis must represent a significant scientific advance.
- If the thesis includes previously published material, permission to reproduce this must be gained from the respective copyright holder.
- They must have been examined and passed during the 12 months prior to nomination.
- Each thesis should include a foreword by the supervisor outlining the significance of its content.
- The theses should have a clearly defined structure including an introduction accessible to scientists not expert in that particular field.

Fredrik Öisjöen

High-T_c SQUIDs for Biomedical Applications: Immunoassays, Magnetoencephalography, and Ultra-Low Field Magnetic Resonance Imaging

Doctoral Thesis accepted by
the Chalmers University of Technology

 Springer

Author
Dr. Fredrik Öisjöen
Department of Microtechnology and
  Nanoscience–MC2
Chalmers University of Technology
Gothenburg
Sweden

Supervisor
Prof. Dr. Alexey Kalabukhov
Department of Microtechnology and
  Nanoscience–MC2
Chalmers University of Technology
Gothenburg
Sweden

ISSN 2190-5053   ISSN 2190-5061   (electronic)
ISBN 978-3-642-43401-3   ISBN 978-3-642-31356-1 (eBook)
DOI 10.1007/978-3-642-31356-1
Springer Heidelberg New York Dordrecht London

Parts of this thesis have been published in the following journal articles:

I. F. Öisjöen, P. Magnelind, A. Kalaboukhov, D. Winkler. "High-T_c SQUID gradiometer system for immunoassays". *Supercond. Sci. Technol.* **2**. 034004 (4pp), 2008.

II. F. Öisjöen, J. F. Schneiderman, M. Zaborowska, S. Karthikeyan, P. Magnelind, A. Kalabukhov, K. Petersson, A. P. Astalan, C. Johansson, D. Winkler. "Fast and sensitive measurement of specific antigen-antibody binding reactions with magnetic nanoparticles and HTS SQUID". *IEEE Trans. Appl. Supercond.* **19**(3). 848–852, 2009.

III. V. Schaller, A. Sanz-Velasco, A. Kalabukhov, J. F. Schneiderman, F. Öisjöen, A. Jesorka, A. P. Astalan, A. Krozer, C. Rusu, P. Enoksson, D.Winkler. "Towards an electrowetting-based digital microfluidic platform for magnetic immunoassays". *Lab on a Chip.* **9**. 3433–3436, 2009.

IV. F. Öisjöen, J. F. Schneiderman, A. P. Astalan, A. Kalabukhov, C. Johansson, D. Winkler. "A new approach for bioassays based on frequency- and time-domain measurements of magnetic nanoparticles". *Biosensors and Bioelectronics.* **25**. 1008–1013, 2010.

V. F. Öisjöen, J. F. Schneiderman, A. P. Astalan, A. Kalabukhov, C. Johansson, D. Winkler. "The need for stable, mono-dispersed, and biofunctional magnetic nanoparticles for one-step immunoassays". *J. Phys: Conf. Ser.* **200**. 122006, 2010. doi:10.1088/1742-6596/200/12/122006.

VI. F. Öisjöen, J. F. Schneiderman, G. A. Figueras, M. L. Chukharkin, A. Kalabukhov, A. Hedström, M. Elam, D. Winkler. "High-T_c superconducting quantum interference device recordings of spontaneous brain activity: Towards high-T_c magnetoencephalography. *Appl. Phys. Lett.* **100**. 132601, 2012. doi:10.1063/1.3698152.

Supervisor's Foreword

It is our great pleasure to introduce Dr. Fredrik Öisjöen's thesis work for the Springer Thesis Prize. This challenging research has focused on developing proof-of-principle measurements with systems that incorporate superconducting technology in the biomedical arena. The activities include two topic areas, both of which are specific to biomedical applications of high critical temperature (high-T_c) superconducting quantum interference devices (SQUIDs). The most significant contributions to the field were:

I. Both theoretically and experimentally demonstrating the efficacy and compatibility of biomolecular detection via rapid time-domain and high-resolution frequency domain SQUID-based recordings of bio-functionalized magnetic nanoparticles.

II. Opening the field of high-T_c SQUID-based magnetoencephalography (MEG) for recording brain function. This was made possible by proving (contrary to the common belief) that a simple and reliable fabrication process could generate high- T_c SQUID detectors sensitive enough to match, or potentially exceed, the signal-to-noise ratios of their conventional low-T_c counterparts when recording magnetic signals generated by brain activity. Remarkably, the manuscript reporting the MEG results demonstrating spontaneous brain signals using high-T_c SQUIDs, was accepted within just 5 working days of submission to Applied Physics Letters.

The work by Fredrik Öisjöen does not simply improve on existing technology but opens possibilities for novel advanced medical devices and techniques. Fast magnetic immunoassays can be used to improve the sensitivity of existing biomolecular detection techniques or create new, unique point-of-care diagnostic systems. High-T_c SQUID technology can possibly improve the clinical availability of MEG systems as well as provide new insights into how our brain works. Combined together with ultra-low field magnetic resonance imaging, this technology can be used to acquire both functional and structural brain images at the same time. Such a capability is critical for improving patient outcomes in a variety of clinical applications. Successful development of three different techniques based on the same sensor technology makes Dr. Öisjöen's thesis work of high scientific quality that is worthy of recognition. We hope the reader will enjoy this very well-written and exciting thesis.

Gothenburg, Sweden, 26 April 2012   Dag Winkler
                                                        Alexey Kalabukhov
                                                     Justin F. Schneiderman

Acknowledgments

First of all I would like to thank my examiner and main supervisor Dag Winkler for giving me the opportunity to work in this exciting field after my diploma work and for guidance and advice leading to this thesis.

I wish to express my gratitude to my co-supervisor, Alexey Kalabukhov, for valuable discussions, for excellent teaching of cleanroom processing and for all the time and effort he has put in to support me.

I am grateful to my co-supervisor Justin Schneiderman for teaching me general experimental techniques, all the support in the laboratory, and for help with articles.

I would like to thank John Clarke at UC Berkeley for allowing me to work in his lab for 3 months. This is where I learned the basics of ULF-MRI in "team Scandinavia" together with Koos Zerenhoven. Thanks also to Sarah Busch and Michael Hatridge.

Many thanks to Andrea Prieto Astalan, Jakob Blomgren, Karolina Petersson, Magdalena Zaborowska, and Christer Johansson at Imego AB for all the collaborative work and valuable discussions regarding magnetic nanoparticles and assay experiments. Special thanks also to Christer for comments on this thesis.

I thank Vincent Schaller and Anke Sanz-Velasco for the efforts on the EWOD platform and I am hopeful about the ongoing and future collaborative efforts.

I would like to thank Mikael Elam and Anders Hedström at the Sahlgrenska hospital for useful input and discussions regarding the MEG experiments and for the highly valuable EEG sessions. I am looking forward to continuing the ongoing efforts. Thanks also to Göran Pegenius for help in setting up the EEG recordings.

Thanks to Maxim Chukharkin for helping out in the laboratory and in the cleanroom.

I am also thankful for the great work by Magnus Jönsson on the ULF-MRI system. Magnus is one of the main contributors in this project.

Thanks to the cleanroom staff of MC2 for helping out with fabrication issues and training, special thanks to Henrik Fredriksen for all his help.

Staffan Pehrson and Lars Jönsson are acknowledged for machining parts for the experimental setups and Ann-Mari Frykestig and Marie Fredriksson for all the administrative support and Jan Jacobsson for technical support.

I thank my office-mate Samuel Lara for inspiring discussions. I would also like to thank all the past and present members of the Quantum Device Physics laboratory at MC2 for creating the nice atmosphere and environment to work in.

I would like to thank my lovely family (Maria, Hans, Johanna, Dan, Sally, and everyone else) and all of my friends for their support and finally, thank you Magdalena for always supporting and encouraging me!

Financial support is acknowledged from the European Commission Framework Program 7 (FP7/2007–2013) project MEGMRI (grant ageement no. 200859), the European Commission Framework Program 6 (FP6/2005–2008) project Biodiagnostics (grant ageement no. NMP4-CT-2005-017002) the Kristina Stenborgs foundation, Knut och Alice Wallenberg fund, the Swedish research council, the Swedish foundation for strategic research, Chalmers friends, and Chalmerska forskningsfonden.

Contents

Symbols and Abbreviations

Symbol	Meaning
α	Exponent for distribution of Néel relaxation times in ac-susceptibility
β	Exponent for distribution of Néel relaxation times in magnetorelaxometry
β_c	Hysteresis parameter
β_L	Inductance parameter
Γ	Noise parameter/gyromagnetic ratio of protons ($2\pi^*42.58$ Hz/μT)
$\Delta f'$	Inhomogeneous linewidth of NMR
f_L	Larmor frequency
η	Viscosity
$\hbar$	Planck's constant
θ	Angle from xy-plane
λ	Wavelength
σ	Geometrical standard deviation
σ_0	Initial standard deviation
τ_{eff}	Effective magnetic relaxation time
τ_B	Brownian relaxation time
τ_N	Néel relaxation time
τ_0	Material specific characteristic relaxation time
τ_e	Measurement time
Φ_0	Flux quantum
Φ	Magnetic flux
Φ_N	Thermally induced flux noise
ϕ	Phase
χ_{0B}	Static susceptibililty due to Brownian relaxation
χ_{0N}	Static susceptibililty due to Néel relaxation
χ	Complex magnetic susceptibility
χ''	Imaginary part of the complex magnetic susceptibility
χ'	Real part of the complex magnetic susceptibility
ω_j	Width of junction
ω	Angular frequency
A	Area

(continued)

(continued)

Symbol	Meaning
A_p	Area of pick up loop
A_{eff}	Effective area of SQUID magnetometer
A_s	Effective area of bare SQUID
B_0	Measurement field for NMR/MRI
$B_{x,y,z}$	Earth's field cancellation fields for NMR/MRI in x, y, z, respectively
B_{p1}	Pre-polarizing field, Helmholtz coil
B_{p2}	Pre-polarizing field, solenoid coil
B	Magnetic field
C	Capacitance
C3/C4	Locations (motor cortex) on the scalp for MEG/EEG
D_S	Shliomis diameter
E	Electric field
E	Energy
e	Electron charge
f	Frequency
f_B	Brownian relaxation frequency
f_m	Measurement frequency
f_J	Josephson frequency
$G_{x,y,z}$	z-Gradient fields for NMR/MRI in x, y, z, respectively
H	Magnetic excitation field
I	Current
I_c	Critical current of junction
I_n	Noise current
I_b	Bias current
I_s	Screening current
J	Current density
J^i	Impressed/primary current density
J_c	Critical current density
K	Magnetic anisotropy constant
k	Coupling constant of SQUID
k_B	Boltzmann constant
L_P	Inductance of pick up loop
L_{sl}	SQUID slit inductance
L_b	SQUID junction bridge inductance
L	SQUID inductance
M_0	Initial magnetization
M_n	Initial strength of magnetization due to Néel relaxation
M	Magnetization
O1/O2	Locations (occipital lobe) on the scalp for MEG/EEG
Q	Current dipole moment
R_s	Shunt resistor
R_d	Dynamic resistance
R_f	Feedback resistor
r_H	Hydrodynamic radius
r_0	Initial hydrodynamic radius

(continued)

(continued)

Symbol	Meaning
R_n	Normal resistance of junction
S_B	Field sensitivity
S_Φ	Power spectral density of SQUID flux noise
S_V	Power spectral density of SQUID voltage noise
T_c	Critical temperature
T	Temperature
T_1	Longitudinal relaxation rate in NMR
T_2	Intrinsic transversal relaxation rate in NMR
T_2'	Extrinsic relaxation rate in NMR
$T_2{}^*$	Total transversal relaxation rate in NMR
t	Time
T_B	Blocking temperature
V_{out}	Output voltage
V	Voltage
V_B	Blocking volume
V_p	Volume of magnetic nanoparticle single domain
V_Φ	SQUID flux to voltage transfer function

Abbreviations

Ac	Alternating current
AC	SAc susceptibility
AFM	Atomic force microscopy
Dc	Direct current
DIW	Distilled water
ECG	Electrocardiography
EEG	Electroencephalography
ELISA	Enzyme linked immunosorbent assay
EIA	Enzyme immunoassay
EWOD	Electrowetting on dielectric
f.c.c	Face-centered cubic
FID	Free induction decay
FLL	Flux locked loop
fMRI	Functional magnetic resonance imaging
FWHM	Full-width-half-maximum
HTS	High temperature superconductor
JJ	Josephson junction
LTS	Low temperature superconductor
MCG	Magnetocardiography
MEG	Magnetoencephalography
MIA	Magnetic immunoassay
MNP	Magnetic nanoparticle
MRI	Magnetic resonance imaging
MRX	Magnetorelaxometry
NDE	Non destructive evaluation
NMR	Nuclear magnetic resonance
PBS	Phosphate buffered saline
PET	Positron emission tomography
PLD	Pulsed laser deposition
PSA	Prostate specific antigen
PSA10	Prostate specific antibody type 10

PSA66	Prostate specific antibody type 66
QCM	Quartz crystal microbalance
RCA	Rolling circle amplification
RCSJ	Resistively and capacitively shunted junction
RIA	Radioimmunoassay
SERF	Spin-exchange relaxation-free
SNR	Signal to noise ratio
SPR	Surface plasmon resonance
SQUID	Superconducting quantum interference device
STO	$SrTiO_3$
ULF	Ultra-low field
UV	Ultraviolet
YBCO	$YBa_2Cu_3O_{7-\delta}$

Chapter 1
Introduction

The superconducting quantum interference device (SQUID) is one of the most sensitive detectors available for measurements of magnetic fields. Due to its unrivaled sensitivity, it has been employed in a variety of applications. One of the most successful application of SQUIDs is for measurements of the tiny magnetic fields produced by the firing neurons in a human brain. This application is known as magnetoencephalography (MEG) and is one of the topics of this thesis. In a state-of-the-art MEG system, a helmet shaped dewar incorporates several hundred SQUID sensors. Before the invention of the SQUID, the existence of magnetic fields produced by neural currents was proven by David Cohen in 1968 by using Faraday type detection with induction coils [1]. Conveniently, a few years after this discovery, the SQUID was invented and one of the first applications was to reproduce what Cohen did, but now the coil was replaced with a SQUID magnetometer [2]. This revolutionized the field of neuroscience, and today the majority of commercial SQUIDs produced worldwide are mounted in MEG helmets. Some important applications of MEG are pre-surgical mapping of brain functions for brain tumor surgery [3] and localization of epileptic foci [4, 5], but also in basic epilepsy and neuroscience research [6].

In the late 1990s, SQUIDs were suggested to be used as biosensors. The idea, first published by Kötitz et al. [7], was to use magnetic nanoparticles (MNPs) as labels in a magnetic immunoassay (MIA). This is the second topic covered in this thesis. The combination of an extremely sensitive magnetometer and magnetic labels could provide an incredibly sensitive tool for diagnostics. The most commonly used method today is the enzyme-linked immunosorbent assay (ELISA) where an enzyme label is used. The enzyme label reacts with an added specific chromogenic substrate and the intensity of light at a certain wavelength is detected. One advantage of using magnetic nanoparticles is that the label is more stable and the assay can be performed with fewer washing steps. Furthermore, it is easier to accomplish a magnetic impurity-free environment than a fluorescent one. The SQUID method can also challenge the conventional methods in terms of sensitivity to an analyte [8–11].

SQUIDs have also been employed as sensors in NMR/MRI systems in ultra-low magnetic fields (ULF-MRI) [12–14] where the feature of frequency independent

F. Öisjöen, *High-T$_c$ SQUIDs for Biomedical Applications: Immunoassays, Magnetoencephalography, and Ultra-Low Field Magnetic Resonance Imaging*, Springer Theses, DOI: 10.1007/978-3-642-31356-1_1, © Springer-Verlag Berlin Heidelberg 2013

magnetic field sensitivity of SQUIDs is crucial. NMR/MRI in ultra-low magnetic fields has advantages related to cost and complexity of the system as well as patient convenience. More importantly, it enables co-registration of MEG and MRI [15] and it has also been shown to enable enhanced contrast of cancerous prostate tissue to healthy prostate tissue (ex-vivo) without using a contrast agent [16]. ULF-MRI is the last topic covered in this thesis and is in our case motivated by the suggested incorporation with MEG. Such a hybrid MEGMRI system may improve the accuracy of MEG reconstructions.

SQUIDs have been used for several applications other than MEG, MRI and MIA. Almost in parallel with the development of MEG, SQUIDs were also applied to measurements of the magnetic fields produced by the human heart. In magnetocardiography (MCG), a multichannel SQUID system records the complex magnetic field pattern produced by the currents flowing in the muscle that makes up the heart. One important applications of MCG is fetal MCG where conventional methods fail (see review by Koch [17]).

Other applications of SQUID sensors are e.g. the SQUID microscope [18], non-destructive evaluation of e.g. bridges [19], and for geophysical applications [20]. For a review of SQUIDs and applications, see [21–23].

Lately, SQUIDs have been challenged by spin-exchange relaxation-free (SERF) atomic magnetometers [24–26]. SERF sensors contain a small chamber with a high density of alkali metal vapor, lasers for optical pumping and detection, and photo diodes. The technique relies on optical absorption of spin-polarized electrons in a small magnetic field. SERF sensors with magnetic field resolutions of $3.5\,\mathrm{fT}/\sqrt{\mathrm{Hz}}$ above $10\,\mathrm{Hz}$ have recently been demonstrated for MEG recordings [25]. A SERF sensor is cryogen-free, however, the vapor cell has to be heated to roughly $200\,°\mathrm{C}$. Although the lack of a cryogen eliminates dewar noise inevitable in a SQUID setup, at present, the low frequency noise (say, below $10\,\mathrm{Hz}$) of SQUIDs is still superior.

Another sensor technology that has emerged is based on a giant magnetoresistance sensor (GMR) with a superconducting pick-up coil, known as a mixed sensor [27]. Such sensors have successfully been applied to ULF-MRI [28]. However, $1/f$-noise remains a problem yet to be solved.

The materials that can be used for fabricating SQUIDs can be categorized as high-T_c and low-T_c superconductors with typical operation temperatures of $77\,\mathrm{K}$ (liquid nitrogen) and $4.2\,\mathrm{K}$ (liquid helium), respectively. In low-T_c SQUID technology, a magnetic field resolution below $1\,\mathrm{fT}/\sqrt{\mathrm{Hz}}$ has been demonstrated [29]. Moreover, low-T_c SQUID magnetometers with magnetic field sensitivities of a few $\mathrm{fT}/\sqrt{\mathrm{Hz}}$ are commercially available [30]. For high-T_c SQUIDs, $3.5\,\mathrm{fT}/\sqrt{\mathrm{Hz}}$ above $10\,\mathrm{Hz}$ [31] is presently among the highest sensitivities achieved. To put these numbers in context, the magnetic field that can be measured by an average SQUID magnetometer is roughly 1 billion (10^9) times weaker than the Earth's magnetic field. The major challenge for SQUID performance in high-T_c technology is to reduce low frequency noise, especially that generated in multilayer structures required for flux transformers.

In the scope of this thesis, only high-T_c is considered. Although the magnetic field sensitivity of the low-T_c SQUIDs is superior to the high-T_c equivalent, there are several important advantages of high-T_c technology. Firstly, liquid nitrogen systems at 77 K require less thermal insulation than liquid helium systems operating at 4.2 K. This advantage leads to a reduction of the spacing between the cold sensor and a room temperature sample such as MNPs or a human scalp, to less than 1 mm compared with 2–4 cm, typical for a liquid helium system. For a magnetic dipole source whose field decays as the inverse cube of the distance from the source, this can be crucial. For example, measurements on microdroplets of MNPs for immunoassays (important for e.g. small sample volumes, parallelization and high throughput) would not be optimized with such a large source-sensor separation. For higher order sources this becomes even more important. This can be the case for magnetic fields produced by complicated patterns of brain activity recorded by MEG. Secondly, a liquid nitrogen system is more flexible than a liquid helium system. For MEG, the rigid helmet dewar that does not fit arbitrary head shapes could be replaced by a flexible array of SQUIDs mounted in novel cooling systems [32, 33]. This way, each individual SQUID sensor could be aligned individually with the scalp. Lastly, the expensive liquid helium can be replaced by cheap liquid nitrogen.

Aim and Outline

The aim of this thesis is to describe the development and demonstrate the feasibility of high-T_c SQUID systems for biomedical applications including magnetic immunoassays (MIAs), magnetoencephalography (MEG), and ultra-low field NMR/MRI (ULF-NMR/MRI). The thesis also describes the development of the high-T_c SQUID sensors themselves.

The development of the MIA system was part of an EU FP6 project, Biodiagnostics, with the aim to develop medical diagnostic tools based on the most sensitive magnetic detectors available. The MEG and ULF-MRI part of the thesis was part of an EU FP7 project, MEGMRI, with the aim to develop a hybrid MEG and ULF-MRI system.

In Chap. 2, the relevant background to superconductivity and SQUIDs are introduced. Details on high-T_c SQUID sensor considerations, fabrication and performance are provided as well as characteristics of the fabricated SQUID sensors. Chapter 3 introduces the field of immunoassays and also the relevant terminology of magnetic nanoparticles. It continues with the experimental details and results of the MIA experiments and the system performance. Chapter 4 includes the relevant background to MEG, the experimental details and results of the MEG recordings and discussion of the results. In Chap. 5, the setup for ULF-NMR/MRI is described along with the preliminary results obtained. Chapter 6 concludes the thesis with a summary and an outlook.

References

1. D. Cohen, Magnetoencephalography: Evidence of magnetic fields produced by alpha-rhythm currents. Science **161**, 784–786 (1968)
2. D. Cohen, Magnetoencephalography: detection of the brain's electrical activity with a superconducting magnetometer. Science **175**, 664–666 (1972)
3. T. Roberts, P. Ferrari, D. Perry, H. Rowley, M. Berger, Presurgical mapping with magnetic source imaging: comparisons with intraoperative findings. Brain Tumor Pathol. **17**, 57–64 (2000)
4. A. Ray, S. Bowyer, Clinical applications of magnetoencephalography in epilepsy. Ann. Indian Acad. Neurol. **13**, 14–22 (2010)
5. H. Stefan, C. Hummel, G. Scheler, A. Genow, K. Druschky, C. Tilz, M. Kaltenhäuser, R. Hopfengärtner, M. Buchfelder, J. Romstöck, Magnetic brain source imaging of focal epileptic activity: a synopsis of 455 cases. Brain **126**(11), 2396–2405 (2003)
6. P.C. Hansen, M.L. Kringelbach, R. Salmelin (eds)., *MEG: An Introduction to Methods* (Oxford University Press, New York, 2010)
7. R. Kötitz, H. Matz, L. Trahms, H. Koch, SQUID based remanence measurements for immunoassays. IEEE Trans. Appl. Supercond. **7**(2), 3678–3681 (1997)
8. C. Yang, S. Yang, J. Chieh, H. Horng, C. Hong, H. Yang, K.H. Chen, B.Y. Shih, T. Chen, M. Chiu, Biofunctionalized magnetic nanoparticles for specifically detecting biomarkers of Alzheimer's disease in vitro. ACS Chem. Neurosci. **2**, 500–505 (2011)
9. K. Enpuku, T. Minotani, M. Hotta, A. Nakahodo, Application of high T_c SQUID magnetometer to biological immunoassays. IEEE Trans. Appl. Supercond. **11**(1), 661–664 (2001)
10. M. Strömberg, J. Göransson, K. Gunnarsson, M. Nilsson, P. Svedlindh, M. Strømme, Sensitive molecular diagnastics using volume-amplified magnetic nanobeads. Nanoletters **8**(3), 816–821 (2008)
11. D. Eberbeck, C. Bergemann, S. Hartwig, U. Steinhoff, L. Trahms, Quantification of specific bindings of biomolecules by magnetorelaxometry. J. Nanobiotechnol. **6**(4), 13 (2008)
12. J. Clarke, M. Hatridge, M. Möble, SQUID-detected magnetic resonance imaging in microtesla fields. Annu. Rev. Biomed. Eng. **9**, 389–413 (2007)
13. V.S. Zotev, A.N. Matlachov, P.L. Volegov, A.V. Urbaitis, M.A. Espy, R.H. Kraus, SQUID-based instrumentation for ultralow-field MRI. Supercond. Sci. Technol. **20**, 367–373 (2007)
14. M. Burghoff, H.H. Albrecht, S. Hartwig, I. Hilschenz, R. Körber, T.S. Thömmes, H.J. Scheer, J. Voigt, L. Trahms, SQUID system for MEG and low field magnetic resonance imaging. Metrol. Meas. Syst. **16**, 371–375 (2009)
15. P.E. Magnelind, J.J. Gomez, A.N. Matlashov, T. Owens, J.H. Sandin, P.L., Volegov, M.A. Espy, Co-registration of interleaved meg and ulf mri using a 7 channel low-T_c system. IEEE. Trans. Appl. Supercond. **21**, 456–460 (2011)
16. S.E. Busch, *Ultra-Low Field MRI of Prostate Cancer Using SQUID Detection*. Ph.D. Thesis, University of California at Berkeley, 2011
17. H. Koch, SQUID magnetocardiography: status and perspectives. IEEE. Trans. Appl. Supercond. **11**, 49–59 (2001)
18. J.R. Kirtley, M.B. Ketchen, K.G. Stawiasz, J.Z. Sun, W.J. Gallagher, S.H. Blanton, S.J. Wind, High-resolution scanning SQUID microscope. Appl. Phys. Lett. **66**(9), 1138–1140 (1995)
19. H.-J. Krause, M. Kreutzbruck, Recent developments in SQUID NDE. Phys. C. **368**, 70–79 (2002)
20. C.P. Foley, K.E. Leslie, R. Binks, C. Lewis, W. Murray, G.J. Sloggett, S. Lam, B. Sankrithyan, N. Savvides, A. Katzaros, K.H. Muller, E.E. Mitchell, J. Pollack, J. Lee, D.L. Dart, R.R. Barrow, M. Asten, A. Maddever, G. Panjkovis, M. Downey, C. Hoffman, R. Turner, Field trials using HTS SQUID magnetometers for ground based and airborne geophysical applications. IEEE. Trans. Appl. Supercond. **9**, 3786–3792 (1999)
21. R. Kleiner, D. Koelle, F. Ludwig, J. Clarke, Superconducting quantum interference devices: state of the art and applications. Proc. IEEE. **92**(10), 1534–1548 (2004)

22. J. Clarke, A.I. Braginski, *The SQUID Handbook*, vol. 1 (WILEY-VCH, Weinheim, 2006)
23. J. Clarke, A.I. Braginski, *The SQUID Handbook*, vol. 2 (WILEY-VCH, Weinheim, 2006)
24. J.C. Allred, R.N. Lyman, T.W. Kornack, M.V. Romalis, High-sensitivity atomic magnetometer unaffected by spin-exchange relaxation. Phys. Rev. Lett. **89**(13), 130801 (2002)
25. H. Xia, B.-A. Baranga, D. Hoffman, M.V. Romalis, Magnetoencephalography with an atomic magnetometer. Appl. Phys. Lett. **89**(21), 104–211 (2006)
26. I.K. Kominis, T.W. Kornack, J.C. Allred, M.V. Romalis, A subfemtotesla multichannel atomic magnetometer. Nature **422**, 596–599 (2003)
27. M. Pannetier, C. Fermon, G. Le Goff, J. Simola, E. Kerr, Femtotesla magnetic field measurement with magnetoresistive sensors. Science **304**, 1648–1650 (2004)
28. N. Sergeeva-Chollet, H. Dyvorne, J. Dabek, Q. Herreros, H. Polovy, G. Le Goff, G. Cannies, M. Pannetier-Lecoeur, C. Fermon, Low field MRI with magnetoresistive mixed sensor. J. Phys. Conf. Ser. **303**(1), 012055 (2011)
29. D. Drung, S. Bechstein, K.P. Franke, M. Scheiner, T. Schurig, Improved direct-coupled dc SQUID read-out electronics with automatic bias voltage tuning. IEEE. Trans. Appl. Supercond. **11**(1), 880–883 (2001)
30. D. Drung, C. Aßmann, J. Beyer, A. Kirste, M. Peters, F. Ruede, T. Schurig, Highly sensitive and easy-to-use SQUID sensors. IEEE. Trans. Appl. Supercond. **17**(2), 699–704 (2007)
31. M.I. Faley, U. Poppe, K. Urban, D.N. Paulson, R.L. Fagaly, A new generation of the HTS multilayer dc-SQUID magnetometers and gradiometers. J. Phys. Conf. Ser. **43**, 1199–1202 (2006)
32. N. Khare, P. Chaudhari, Operation of bicrystal junction high-T_c direct current-SQUID in a portable microcooler. Appl. Phys. Lett. **65**, 2353–2355 (1994)
33. P.P.P.M. Lerou, H.J.M. ter Brake, J.F. Burger, H.J. Holland, H. Rogalla, Characterization of micromachined cryogenic coolers. J. Micromech. Microeng. **17**, 1956–1960 (2007)

Chapter 2
High-T_c SQUIDs

Superconductivity was discovered by Kamerlingh Onnes in 1911 after he succeeded in liquefying helium [1]. Kamerlingh Onnes discovered that the electrical resistance of mercury vanished when it was cooled to below 4.2 K. The transition temperature of a superconductor is known as the critical temperature, T_c, and is a material-specific property. Zero resistance is one of the characteristic properties of superconductors and the other defining property is called the Meissner effect, discovered in 1933 [2]. The characteristic of the Meissner effects is that magnetic fields are completely expelled from inside the bulk of a superconductor when cooled to below T_c. The electrons in a superconductor pair up in so-called Cooper pairs [3] due to interaction with lattice vibrations, known as phonons, that make it possible for electrons to travel through the material without dissipation.

An important breakthrough in superconducting technology came in 1986 when Bednorz and Müller discovered the first high temperature superconductor with a T_c of 35 K [4] for which they were awarded the 1987 Nobel prize in physics. Shortly after, superconductivity in $YBa_2Cu_3O_{7-\delta}$ (YBCO) was discovered [5] with $T_c = 93$ K which made it possible to use liquid nitrogen as the refrigerant. YBCO is one of the most commonly used high-T_c superconductors today and is the material used throughout this work.

The discovery of superconductivity and the high-T_c materials opened up the field for new applications. This chapter describes the development of high-T_c SQUIDs including theory, considerations, and fabrication of SQUID sensors.

For more details on superconductivity and devices see for example [6–9].

2.1 The Josephson Effect

A Josephson junction consists of two superconductors separated by a thin insulating barrier. The insulator forms a weak link that Cooper pairs can tunnel through, thus, a current can flow through the barrier without dissipation. The behavior of such a device

F. Öisjöen, *High-T_c SQUIDs for Biomedical Applications: Immunoassays,*
Magnetoencephalography, and Ultra-Low Field Magnetic Resonance Imaging, Springer Theses,
DOI: 10.1007/978-3-642-31356-1_2, © Springer-Verlag Berlin Heidelberg 2013

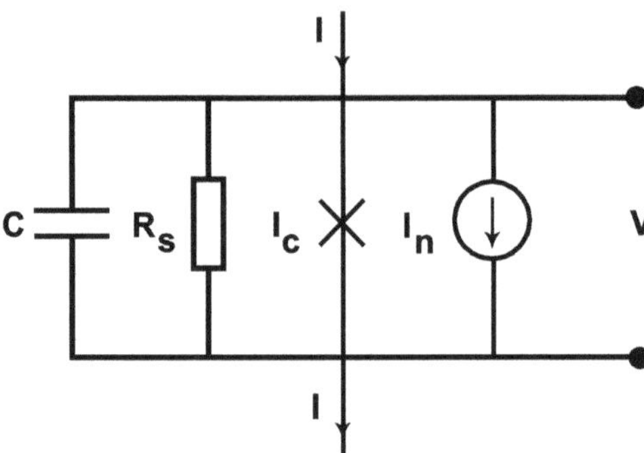

Fig. 2.1 An equivalent circuit of a capacitively (C) and resistively (R_s) shunted Josephson junction with critical current I_c and noise current I_n

was predicted by Josephson in 1962 [10] and the governing Josephson relations are:

$$I = I_c \sin \phi \tag{2.1}$$

$$V = \frac{\hbar}{2e} \frac{\partial \phi}{\partial t} = \frac{\Phi_0}{2\pi} \frac{\partial \phi}{\partial t} \tag{2.2}$$

where I_c is the critical current (the maximum supercurrent that can flow in the junction), ϕ is the phase difference across the junction, $\Phi_0 = h/2e = 2 \cdot 10^{-15}$ Wb is the flux quantum and e is the electron charge. These two equations (Eqs. 2.1 and 2.2) are referred to as the dc and ac Josephson effect, respectively.

A common way of modeling a Josephson junction is by capacitively (C) and resistively (R_s) shunting an ideal junction. This model is called the resistively and capacitively shunted junction (RCSJ) model and is shown in Fig. 2.1 [8, 11, 12]. The current flowing through the model circuit using the Josephson relations can be described by

$$I = C \frac{\Phi_0}{2\pi} \ddot{\phi} + \frac{1}{R_s} \frac{\Phi_0}{2\pi} \dot{\phi} + I_c \sin \phi + I_n, \tag{2.3}$$

where I_n is the noise current from the shunting resistor R_s. From this model arises the Stuart-McCumber parameter:

$$\beta_c = \frac{2\pi I_c R_s^2 C}{\Phi_0}, \tag{2.4}$$

that is used as a condition for hysteresis in the junction's IV characteristics. For highly dissipative junctions used for sensors, $\beta_c < 1$.

In order to retain coupling between the two superconducting electrodes in the presence of thermal noise, the Josephson coupling energy, $I_c \Phi_0 / 2\pi$, has to be larger than the thermal noise [8, 13]. This condition is expressed as

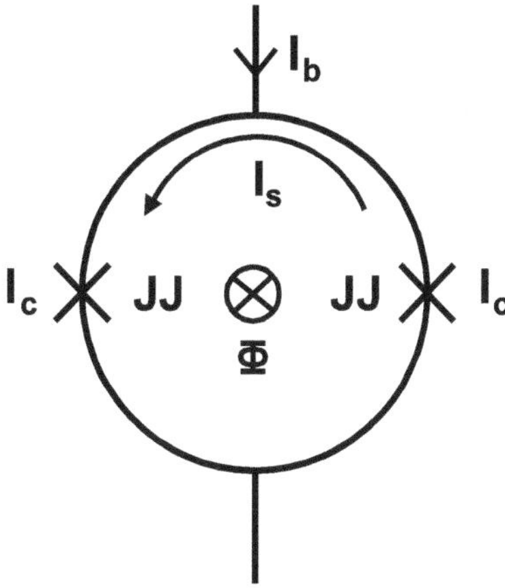

Fig. 2.2 Schematic drawing of a dc SQUID. Two Josephson junctions (JJ) interrupt a superconducting ring. An applied magnetic flux (Φ) directed into the plane of the SQUID induces a screening current (I_s) in the ring. The screening current adds to the bias current (I_b) and alters the critical current of the junctions

$$\Gamma = \frac{2\pi k_B T}{I_c \Phi_0} \leq 1,\qquad(2.5)$$

where T is the operation temperature of the junction.

2.2 The dc SQUID

A superconducting quantum interference device (SQUID) is a flux to voltage transducer and is one of the most sensitive devices for detecting magnetic fields known today. A dc SQUID is a superconducting ring interrupted by two Josephson junctions in parallel as illustrated in Fig. 2.2. An important physical effect (in a SQUID) is that the flux penetrating a superconducting ring is quantized ($\Phi = n\Phi_0, n = 1, 2, 3 \ldots$ [7]). A typical IV-curve of a SQUID with external magnetic flux is shown in Fig. 2.3. When the SQUID is biased with a current bias I_b slightly above the critical current of the SQUID ($2I_c$) the junctions are in the resistive state. Since the critical current of the junctions is modulated with the applied magnetic flux, the output voltage will also be a function of magnetic flux as illustrated in Fig. 2.3.

The IV-curve and the operation of a SQUID can be understood from the following description. If the critical currents of the two junctions are equal, as shown in Fig. 2.2, then the bias current, I_b, will split equally in the superconducting ring. An applied magnetic flux (Φ) induces a circulating current in the ring (with inductance L) given by $I_s = -\Phi/L$. The screening current adds to the bias current through the junctions which means the critical current, as measured, decreases. As the applied flux reaches $\Phi = \Phi_0/2$ the SQUID switches into normal state and lets one flux quantum into the ring, which means I_s changes sign. At this point we are at the minimum critical current of the SQUID as indicated by the lower IV-curve in Fig. 2.3. As the flux increases to one flux quantum, the screening current decreases and the maximum

Fig. 2.3 Measured IV-curve with varied applied flux. The applied flux reduces the critical current and consequently, a voltage modulation appears as a function of the applied magnetic flux (inset)

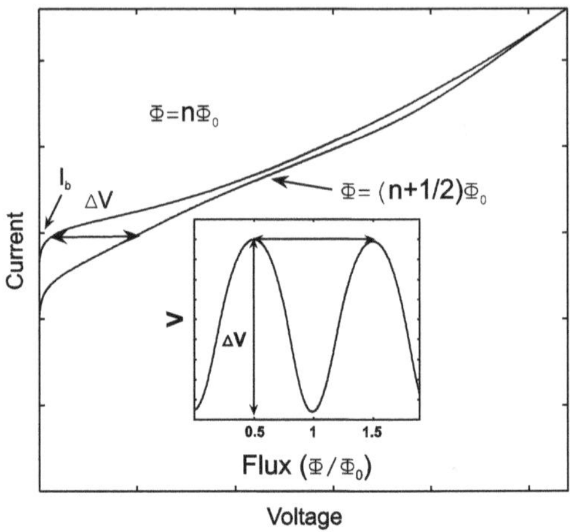

critical current is restored (upper IV-curve in Fig. 2.3). If the SQUID is biased at a constant current (I_b in Fig. 2.3), this yields a voltage modulation as illustrated in the inset of Fig. 2.3.

2.2.1 Noise in dc SQUIDs

There are several types of voltage noise in SQUIDs: Nyquist noise of the Josephson junctions, thermal fluctuations of the critical current and $1/f$ flicker noise associated with motion of flux vortices in the superconducting thin film. The contribution of thermal noise in SQUIDs is more significant in high-T_c devices operating at 77 K than the low-T_c equivalent operating at liquid helium temperatures (4.2 K). If the spectral density of the voltage noise across the SQUID is S_V then the flux noise (typically quoted for SQUIDs) is given by

$$S_\Phi = \frac{S_V}{V_\Phi^2} \tag{2.6}$$

where $V_\Phi = \partial V/\partial \Phi$ is the transfer function of the SQUID. The voltage noise is generated by Nyquist noise from the dynamic resistance in the SQUID at the measurement frequency, f_m, but it is also mixed down to the measurement frequency from $f_J \pm f_m$ (where $f_J = V/\Phi_0 = 483.6\,\text{MHz}/\mu\text{V}$ is the Josephson frequency). The spectral density of the voltage noise (in the limit $\beta_c \ll 1$, $I > I_c$ and $f_m \ll f_J$) is given by [14–16]

$$S_V = \left[1 + \frac{1}{2}\left(\frac{I_c}{I_b}\right)^2\right]\frac{4k_BTR_d^2}{R_s} \tag{2.7}$$

where the first term is the Nyquist noise generated in the dynamic resistance, R_d, and the second term represents the mixed down noise. A second effect of thermal noise is rounding of the IV-curve at low voltages due to current fluctuations [17].

Thermal noise gives several constraints on the SQUID parameters. The effect of thermal noise for coupling between the superconducting electrodes in the Josephson junctions was briefly discussed in Sect. 2.1. Computer simulations by Clarke and Koch [18] suggested an additional factor of 5, thus:

$$\Gamma = \frac{2\pi k_B T}{I_c \Phi_0} \leq 0.2. \tag{2.8}$$

For $T = 77\,\mathrm{K}$, we obtain $I_c > 17\,\mu\mathrm{A}$. For high-T_c SQUIDs one should also keep in mind that the thermally induced flux noise in the SQUID loop, $< \Phi_N^2 >^{1/2} = (k_B T L)^{1/2}$, should be less than one flux quantum which limits the SQUID inductance [8].

The noise in the low frequency regime is usually refered to as $1/f$ noise or flicker noise. The inverse proportionality to frequency makes reduction of flicker noise important for applications where signals are measured at low frequencies, e.g. in MEG. In general, there are two sources of flicker noise. The first one comes from critical current fluctuations due to electrons getting trapped on defects in the barrier and subsequently released [19–21]. The motion of electrons as they trap and release increases the current density locally in the junction and the result is switching of the critical current randomly back and forth as the electrons are released.

The fluctuations of the critical current in the two junctions can be in-phase or out-of-phase. The in-phase component appears as voltage noise in the $V\Phi$-curve shifting the curve in voltage whereas the out-of-phase component gives flux noise and thus a phase shift in the $V\Phi$-curve. By using an appropriate bias scheme, both the in-phase and the out-of-phase component can be reduced substantially [22].

The other type of flicker noise arises from trapped flux that hops between different pinning sites in the superconducting film. Flux gets trapped in the superconducting film when cooled through the transition temperature in the presence of an external magnetic field and the effect is enhanced if the magnetic field is increased. Flux hopping can be reduced by pattering holes in the superconducting film preventing the flux from hopping between pinning sites [23].

2.2.2 SQUID Readout

The SQUID is operated in a flux-locked loop (FLL). The signal from the SQUID is amplified and returned to a feedback coil inductively coupled to the SQUID as shown in Fig. 2.4. The feedback coil generates a flux in the SQUID which opposes the external flux in the SQUID loop. The SQUID is flux locked where the voltage to flux modulation is the steepest in order to achieve highest sensitivity. The signal

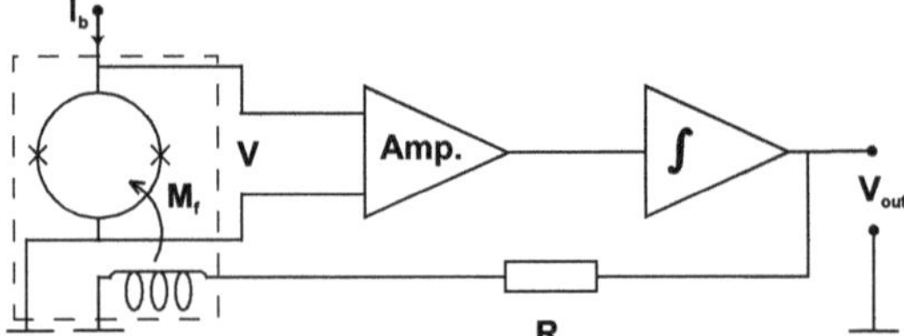

Fig. 2.4 Basic drawing of the FLL SQUID readout. The signal from the SQUID is amplified, integrated and sent back to a feedback coil inductively coupled to the SQUID. The feedback coil cancels the flux in the SQUID. Cold parts are indicated by the *dashed box*

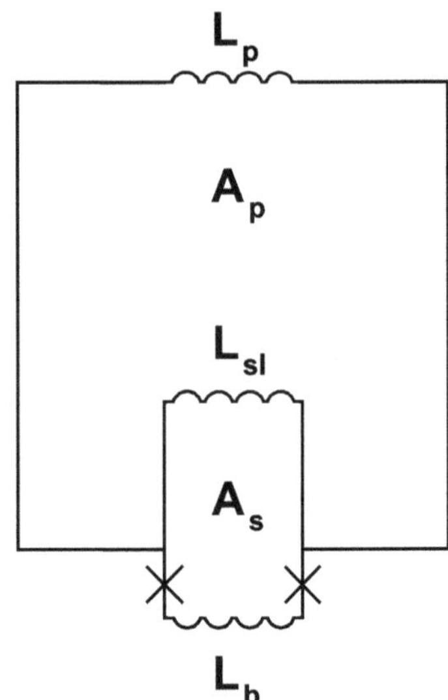

Fig. 2.5 A schematic circuit of a directly coupled magnetometer indicating the inductances of the pick-up coil (L_p), the SQUID slit (L_{sl}), and the junction bridge (L_b). The effective area of the SQUID and the pick-up loop are A_s and A_p respectively. The Josephson junctions are indicated by crosses

fed back to the feedback coil also passes through a feedback resistor (R_f) in order to read out the signal (V_{out}).

The critical current fluctuations giving in-phase and out-of-phase fluctuations are effectively canceled by using a flux-modulated bias-reversal scheme. Bias reversal switches the polarity of the $V\Phi$-characteristics which effectively cancels the in- and out-of-phase fluctuations while remaining sensitive to changes in the $V\Phi$-curve due to applied magnetic flux [22].

2.2.3 SQUID Considerations

Optimization of a SQUID concerns noise, field tolerance and effective area. From simulations it was concluded that $\Gamma = \frac{2\pi k_B T}{I_c \Phi_0} \leq 0.2$ and $\beta_L = 2L I_c/\Phi_0 = 1$ [18, 24]. This puts constraints on the critical current and the SQUID inductance,

$I_c > 17\,\mu\text{A}$ and $L < 59\,\text{pH}$. Simulations done by Enpuku et al. show the dependence of inductance on the noise properties of high-T_c dc SQUIDs [25] and the effect of thermal noise on the voltage modulation [26]. It was shown that the flux noise ($S_\Phi^{1/2}$) scales with the SQUID inductance for $L < 100\,\text{pH}$ and increases substantially for higher inductances.

The effective area of a directly coupled SQUID magnetometer is [27]:

$$A_{eff} = A_s + k\frac{A_p}{L_p}L \tag{2.9}$$

where $k \approx L_{sl}/L$ is a coupling constant, $L = L_{sl} + L_b$ is the SQUID inductance where L_{sl} and L_b are the SQUID slit inductance and the inductance related to the junction bridge respectively, A_s is the effective area of the bare SQUID, A_p and L_p are the effective area and inductance of the pick-up loop respectively (see Fig. 2.5). The effective area of the bare SQUID can be neglected and inserting the coupling constant gives $A_{eff} \propto A_p(L_{sl}/L_p)$. This is also reflected in the noise performance, using $V \approx R/L$ and $S_V = 16k_BTR$ [8] one obtains for the field sensitivity [28]:

$$S_B^{1/2} = \frac{S_V^{1/2}}{V_\Phi A_{eff}} \propto \frac{L}{L_{sl}}. \tag{2.10}$$

In summary, the effective area should be maximized and according to Eq. 2.10 one should minimize L/L_{sl} which was more closely investigated in [28]. The junctions should be designed such that the critical current is higher than $17\,\mu\text{A}$ and the inductance of the SQUID should not be too large ($<59\,\text{H}$). Lines in the film should be kept small in order to avoid flux hopping that causes low frequency noise [23]. Finally, the width of the junctions should be kept small enough not to reduce the critical current modulation due to coupling to magnetic fields. The expression [29]

$$\Delta B \approx \Phi_0\frac{1.84}{\omega_j^2}, \tag{2.11}$$

where ΔB is the maximum magnetic field, and ω_j is the width of the Josephson junction can be used to estimate the maximum junction width given the amplitude of the magnetic field.

2.2.4 Device Layout

Depending on the application and experimental setup a different type of sensor layout may be needed. In the setup for MIA the SQUID sensor was operated without magnetic shielding. In order to reduce the influence of far distant sources of noise one can use a gradiometer layout. Furthermore, for MIA an excitation magnetic

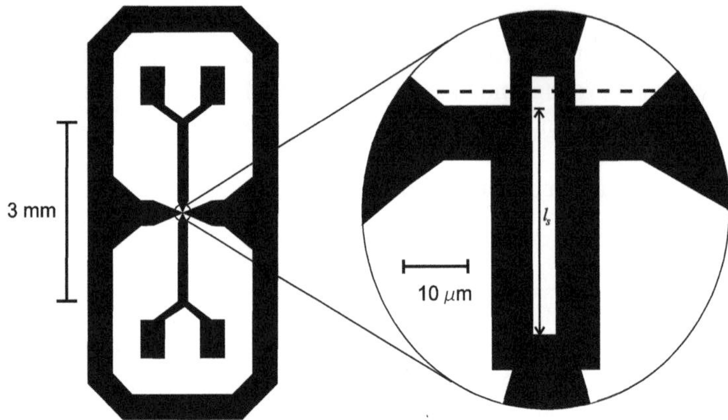

Fig. 2.6 On the *left* is the layout of GRAD1. The two pick-up loops, with width 400 μm, constitute a "figure eight". An homogeneous applied magnetic field will induce circulating currents flowing in the same direction in the two loops. When the currents meet in the middle they cancel, thus the SQUID is blind to homogeneous magnetic fields. The baseline is 3 mm as indicated, thus, the design is optimized for small sample sizes. On the *right* is the layout of the hairpin SQUID in this device. The dashed line indicates the grain boundary of the substrate creating the Josephson junctions in the superconducting ring. The SQUID slit is indicated by the length $l_s = 30\,$μm. The width of the slit and striplines of the SQUID are 3 and 7 μm, respectively

field was applied (see Chap. 3) using a Helmholtz coil that produced a homogeneous magnetic field in the vicinity of the SQUID. Due to these circumstances, gradiometers were more appropriate for this application. Gradiometers were also developed for the ULF-MRI setup, although the results obtained in this thesis on ULF-NMR/MRI were measured with a shielded SQUID magnetometer.

The layout of one of the fabricated SQUID gradiometers (GRAD1) is shown in Fig. 2.6. Currents induced by homogeneous magnetic fields are cancelled in the center of the "figure eight" configuration of the gradiometer since the induced currents will effectively only appear in the perimeter. The SQUID is a so-called hairpin SQUID. It has a thin SQUID slit as opposed to the typical square washer type [30].

The layout of GRAD2-5 (shown in Fig. 2.7) is similar to GRAD1 but the baseline is increased to 4 mm and the SQUID inductance is higher (60 pH, length of slit, width of slit, and width of striplines were 50, 3, and 4 μm, respectively).

GRAD6 is a single gradiometer on a $10\times10\,$mm^2 STO substrate. The layout, shown in Fig. 2.8 incorporates an array of 20 Josephson junctions with widths of 6 μm to serve as flux dams in order to protect the SQUID from the high currents induced during magnetic pulses (for ULF-NMR/MRI, see Chap. 5), but also to reduce the risk of trapping magnetic flux in the thin film close to the SQUID. In addition, the linewidth of the gradiometer was reduced to 200 μm. The area of the pick-up loops was increased in this design.

For the MEG recordings, we used a magnetometer layout since the measurements were made inside a magnetically shielded room and we wanted to maximize the

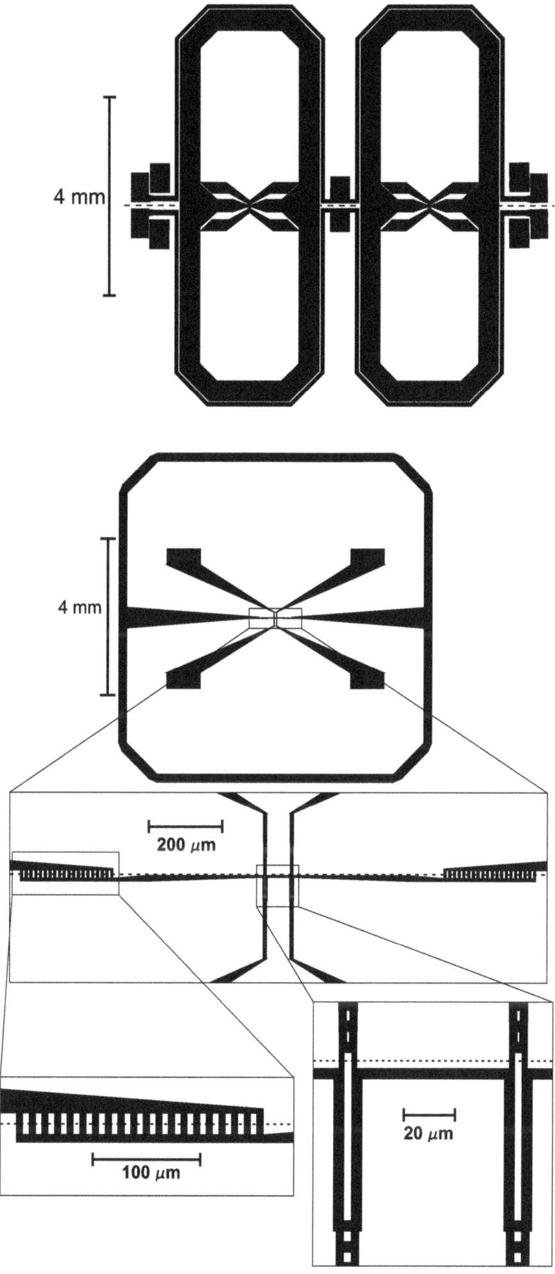

Fig. 2.7 Layout of GRAD2-5. In this design, two gradiometers are fabricated on the same chip (GRAD2-3 and GRAD4-5). The baseline is increased to 4 mm and the SQUID, which is a hairpin type similar to GRAD1 (Fig. 2.6), has length of slit, width of slit, and width of striplines of 50, 3, and 4 μm, respectively. The *dashed line* indicates the grain boundary of the substrate

Fig. 2.8 The layout of GRAD6 shown at different scales. The *uppermost* figure shows the overall layout of the gradiometer. The middle panel shows the flux dam configuration with the two SQUIDs. The bottom left panel shows the 20 Josephson junctions that constitute the flux dam on the left side of the two SQUIDs and the bottom right panel shows the two redundant hairpin SQUIDs. The grain boundary is indicated by the *dashed line*

field sensitivity by making the area of the pick-up loop large. The layout of the magnetometer device is shown in Fig. 2.9. It has a $8 \times 8\,\text{mm}^2$ pick-up loop with a linewidth of 400 μm. For redundancy and higher yield purposes, two SQUIDs were included and placed in series. The hairpin SQUIDs had the same design as in

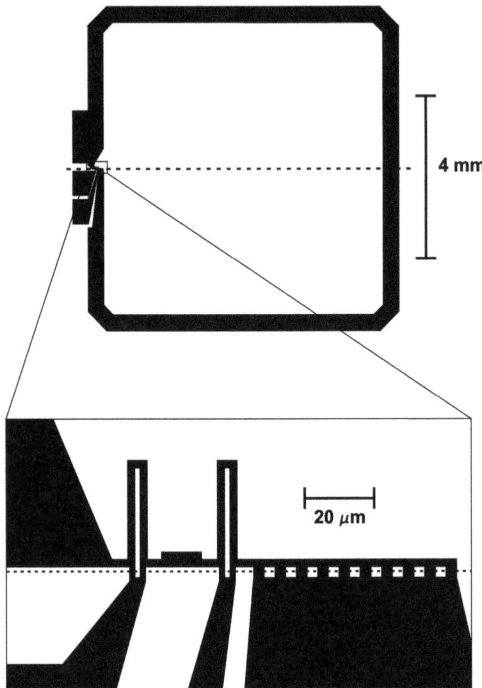

Fig. 2.9 Layout of the SQUID magnetometers. The large pick-up loop is 8×8 mm^2 and incorporates flux dams shown in the zoomed up panel along with the two redundant hairpin SQUIDs. The *dashed line* indicates the grain boundary of the substrate

Table 2.1 A summary of the SQUID devices

Device	Baseline (mm)	Substrate(mm^3 STO)	Seed-layer	SQUID inductance(pH)
GRAD1	3	24°, $5 \times 10 \times 1$	–	46
GRAD2	4	24°, $10 \times 10 \times 1$	–	60
GRAD3	4	24°, $10 \times 10 \times 1$	–	60
GRAD4	4	24°, $10 \times 10 \times 1$	50 nm CeO$_2$	60
GRAD5	4	24°, $10 \times 10 \times 1$	50 nm CeO$_2$	60
GRAD6	4	24°, $10 \times 10 \times 1$	22 nm CeO$_2$	60
MAG1	–	24°, $10 \times 10 \times 1$	–	60
MAG2	–	30°, $10 \times 10 \times 0.5$	22 nm CeO$_2$	60
MAG3	–	30°, $10 \times 10 \times 0.5$	22 nm CeO$_2$	60
MAG4	–	24°, $10 \times 10 \times 1$	22 nm CeO$_2$	60
MAG5	–	24°, $10 \times 10 \times 0.5$	22 nm CeO$_2$	60

GRAD2-6. In order to limit the induced supercurrent in the pick-up loop, arrays of 6 μm Josephson junctions were incorporated in the pick-up loop where the thin film inevitably crosses the grain boundary. Furthermore, this prevents flux hopping in the

Fig. 2.10 A schematic drawing of a pulsed laser deposition system. The laser beam is defined by aperture before it is focused by a lens and enters the chamber where it hits a target. A plume is created by the ejected material from the target. The ejected material is deposited onto the heated substrate

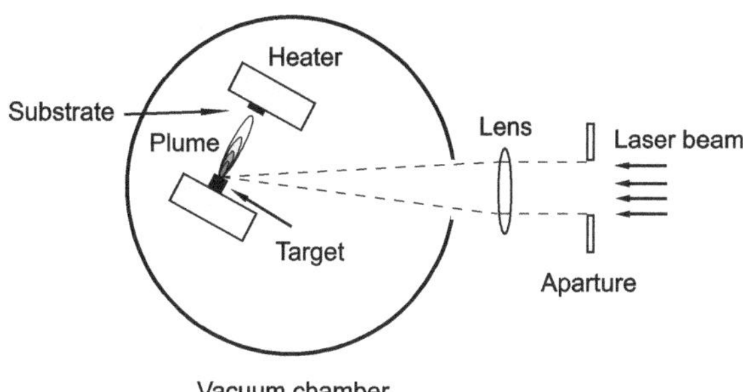

weak parts on the grain boundary of the thin film, which reduces $1/f$ noise. A SQUID magnetometer was also applied for ULF-NMR/MRI (see Chap. 5). A summary of the devices can be found in Table 2.1. The SQUID inductances were extracted from simulations using the 3D-MLSI software [28, 31].

2.3 Device Fabrication

There are several types of practical high-T_c Josephson junctions but in this work YBCO bicrystal grain boundary type junctions have been used. Grain boundary Josephson junctions include, apart from bicrystalline junctions, biepitaxial and step-edge junctions [8, 32]. Examples of other types of junctions are constriction type, ion irradiation and ramp-edge type junctions [8, 33].

For the devices fabricated within this work, grain boundary bicrystal junctions have been utilized due to the reproducibility and simplicity to realize junctions. The bicrystal substrate is made from two $SrTiO_3$ (STO) single crystals bonded together at a misorientation angle of e.g. 24° or 30°. When an epitaxial thin film is deposited onto the substrate the grain boundary is transferred to the thin film creating a weak link with the same misorientation angle. Typical critical current densities (J_c) are on the order of 10^4–10^5 A/cm^2 for YBCO bicrystal junctions with a misorientation angle of 24° [32]. In most of the devices, as shown in Table 2.1, a seed-layer of CeO_2 was sputtered on the substrate before the YBCO thin film was grown. The seed layer is used to improve the quality of the YBCO thin film in the vicinity of the grain boundary of the substrate.

2.3.1 Thin Film Deposition

Deposition of YBCO thin films was made using pulsed laser deposition (PLD) which is a well established technique for growing high-T_c superconducting thin films. The main advantage of PLD compared to other methods is the preservation of the stoichiometry (film composition) which is achieved by the short laser pulses used for target ablation. A disadvantage is formation of large clusters on the film surface. A review on PLD for YBCO deposition can be found in [34].

A schematic of a PLD system is shown in Fig. 2.10. A laser beam is defined by an aperture and focused by a lens before it enters the vacuum chamber. The laser pulses hit a polycrystalline YBCO target inside the chamber. A plasma plume is created by the ejected material from the target. The substrate is placed near the plume which enables the ejected material to be deposited on it.

The UHV-PLD system (DCA instruments) used for growing the thin films of YBCO for our devices consists of a Kr/F excimer laser from Lambda Physik with a wavelength of 248 nm and pulse duration of 20 ns, and a vacuum chamber containing a holder for the target and the substrate. The substrate holder can be set to rotate and/or scan during deposition in order to accomplish thin films on larger substrates (up to 2″) with homogeneous thickness. Heating is done via radiation from a SiC element. This type of heating gives a more homogeneous temperature of the sample compared to thermal heating where the sample is glued onto a heater. The PLD chamber is one part of a cluster system with a common transfer and loading system that enables deposition of oxides and metals in situ.

During thin film growth the substrate is located 51 mm from the target and is heated to around 810 °C. The oxygen pressure during deposition is typically kept at 0.6–0.7 mbar. Finally, after the growth, the film is annealed in oxygen at 500–600 °C for up to 120 min before it is cooled with a rate of 10 °C/min. Thin films with thickness' of 200–350 nm and with T_c's of about 90 K were grown for the devices described. After deposition, the YBCO is protected by a thin layer (∼20 nm) of Au sputtered in situ.

The quality of YBCO thin films is of outmost importance for device fabrication [13]. It is desirable to achieve a high superconducting transition temperature and smooth films, however, they are typically complementary.

Transition Temperature

The transition temperatures of the thin films were determined by measuring the susceptibility of the superconducting film as a function of temperature. The setup consists of two coils with a small separation where the sample is placed. The mutual inductance of the coils is measured as the sample is cooled. At the transition temperature, the susceptibility of the superconductor changes and the mutual inductance between the two coils is altered due to screening currents in the superconductor.

The shape of the transition provides information about the quality of the thin film. It is desirable to achieve a high transition temperature with a narrow transition for optimum operation at 77 K. A representative transition temperature measurement of a YBCO thin film is shown in Fig. 2.11.

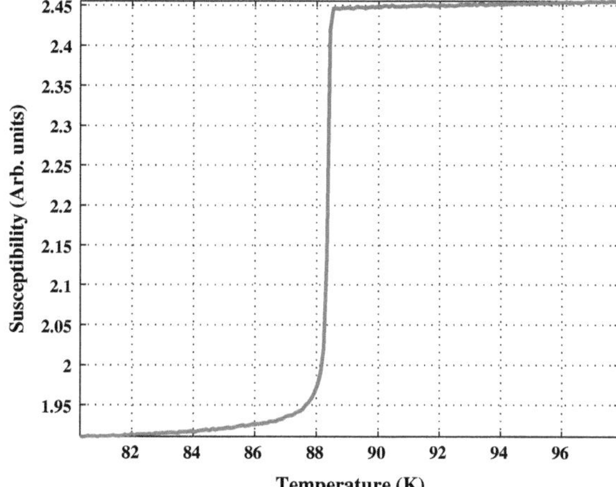

Fig. 2.11 Transition temperature measurement of a YBCO thin film deposited with PLD. The system measures the susceptibility (not calibrated, hence the arbitrary units) of the thin film as it is cooled to below its transition temperature

Fig. 2.12 A microscope photograph of a smooth YBCO thin film with few particles

Surface Morphology

The surface morphology of the thin films was investigated using an optical microscope. This method gives an indication of the surface quality in terms of particle density and smoothness and was sufficient for our purposes. A more thorough investigation of the surface morphology requires atomic force microscopy (AFM). Two optical microscope photographs of YBCO thin films on STO substrates are shown in Figs. 2.12 and 2.13. The YBCO thin film in Fig. 2.12 is an example of a smooth film with few particles whereas the thin film in Fig. 2.13 has more particles and is rougher.

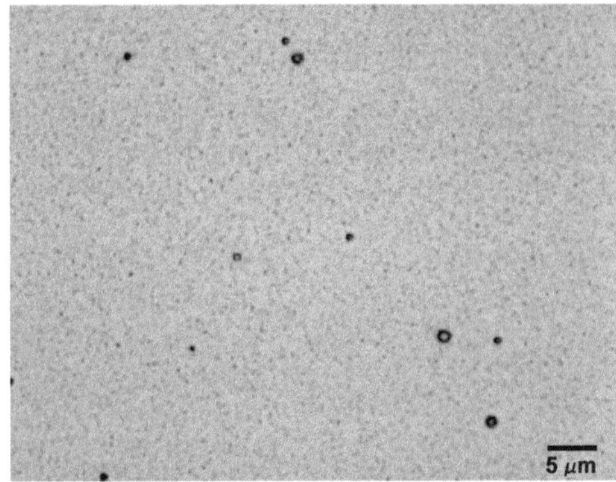

Fig. 2.13 A microscope photograph of a rough YBCO thin film with some particles

2.3.2 Thin Film Processing

Patterning of the YBCO thin film was made with photolithography. A thin layer of UV sensitive photoresist (Shipley S1813) was spun onto the thin film. The photoresist was exposed to UV light ($\lambda = 400\,\text{nm}$) through a chromium mask followed by development in MF-319. Argon ion milling was then used to etch the YBCO. The beam energy during etching was kept between 350 and 500 eV, and the current density to about 0.1–$0.2\,\text{mA/cm}^2$ in order to minimize the damage to the YBCO [13]. Gold pads were patterned using a lift-off process where windows in a layer of photoresist (S1813) were opened with photolithography. Under the S1813, a layer of LOR3A was spun and was used to achieve an undercut in order to facilitate the lift-off done in acetone and ultra-sound bath. The fabrication process is shown in a step-by-step manner in Fig. 2.14. A microscope photograph of GRAD1 is shown in Fig. 2.15.

2.4 SQUID Performance

The SQUID gradiometers were characterized in a shielded environment using a superconducting shield (Bismuth strontium calcium copper oxide, BSCCO) and mu-metal shielding but also unshielded which was particularly important for the immunoassay application. The magnetometers were characterized with a mu-metal shield and a superconducting shield or in the MEG setup inside a magnetically shielded room. For characterization of a SQUID it was glued onto a circuit-board and electrically connected with gold wire-bonds. In this dip-stick configuration, the SQUID was immersed in liquid nitrogen (77 K) either with a superconducting shield inserted into a copper tube covering the SQUID or with only the copper.

One of the fabricated gradiometers (GRAD1, Fig. 2.15) is presented in Paper I. The baseline of the gradiometer was 3 mm and the SQUID inductance was 46 pH. The SQUID on this 5×10 mm^2 large STO chip with a 24° misorientation angle had 3 μm

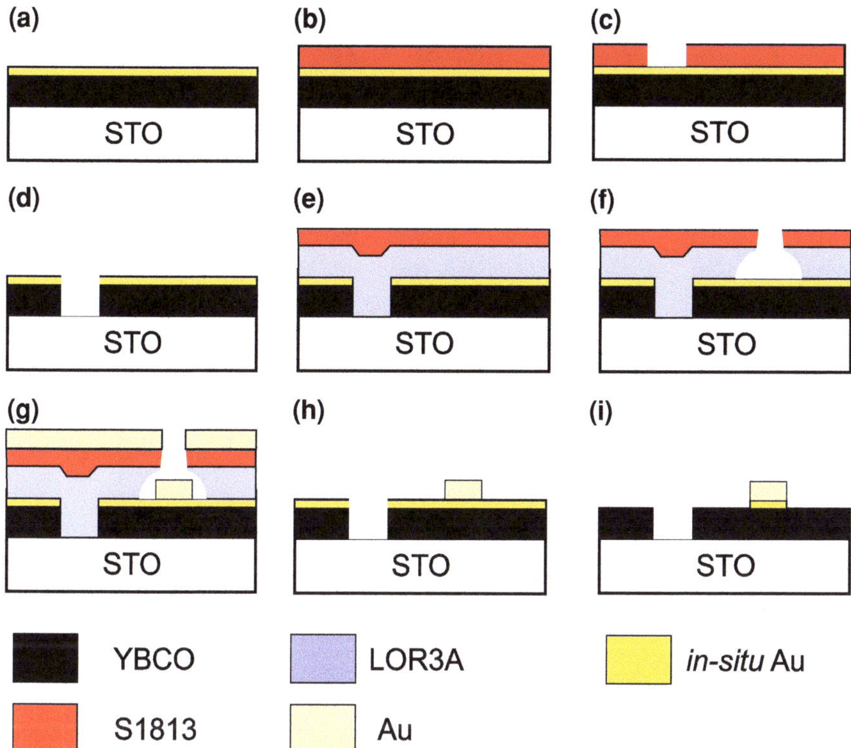

Fig. 2.14 A step-by-step description of the fabrication procedure (the figure is not to scale). **a** First, a YBCO thin film with thickness 200–350 nm is deposited onto a STO bicrystal substrate with an in-situ cap layer of Au ($\sim$20 nm). **b** A layer of photoresist (S1813) is spun onto the sample followed by baking on a hot plate (100°C for 1 min). **c** The photoresist is exposed by UV light through a chromium mask and developed in MF-319 followed by O_2 plasma etching. **d** Ar ion milling for about 120 min is used to etch the YBCO. **e** The next step is to fabricate gold pads which is done with a lift-off process. First, a layer of LOR3A is spun onto the sample followed by baking. Subsequently, a layer of photoresist (S1813) is applied to the sample. After exposure, the photoresist is developed in MF-319 followed by O_2 ashing. **f** Due to different development time of S1813 and LOR3A an undercut is achieved. **g**–**h** Sputtering of gold is followed by lift-off in acetone and ultra-sound bath. **i** Finally, the in-situ gold is removed with a few minutes of Ar ion milling

junctions with normal resistance, $R_n/2 = 1.3\,\Omega$ and critical current $2I_c = 125\,\mu\text{A}$ giving $I_cR_n = 170\,\mu\text{V}$ at 77 K. The flux noise of GRAD1 is shown in Fig. 2.16 measured inside the superconducting shield with two different bias modes (dc and bias reversal). The flux noise at the white level was $4.6\,\mu\Phi_0/\sqrt{\text{Hz}}$ and is comparable with the best achieved flux noise levels for high-T_c SQUIDs [36–40]. The well-balanced gradiometric configuration made operation in unshielded environments stable.

Two other gradiometers (GRAD2-3) had slightly larger baselines (4 mm) and the SQUID inductance was higher (60 pH). Flux modulated IV-curves of GRAD3 are

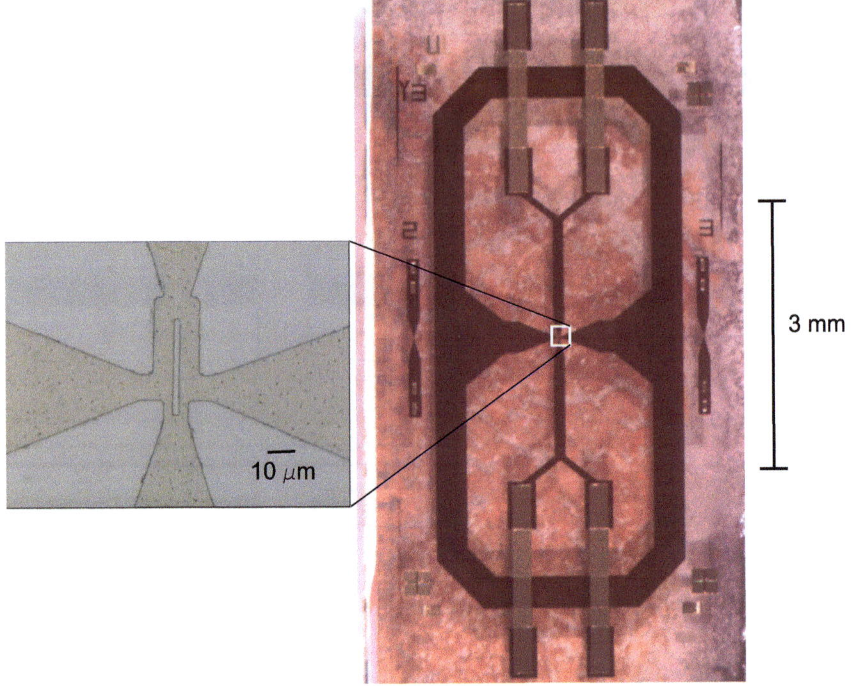

Fig. 2.15 Microscope photograph of GRAD1 and a close up of the hairpin SQUID. Adapted from [35]

Fig. 2.16 Flux noise of GRAD1 measured with shielding using bias reversal and dc bias. As can be seen, the low frequency noise is clearly suppressed by using bias reversal. Adapted from [35]

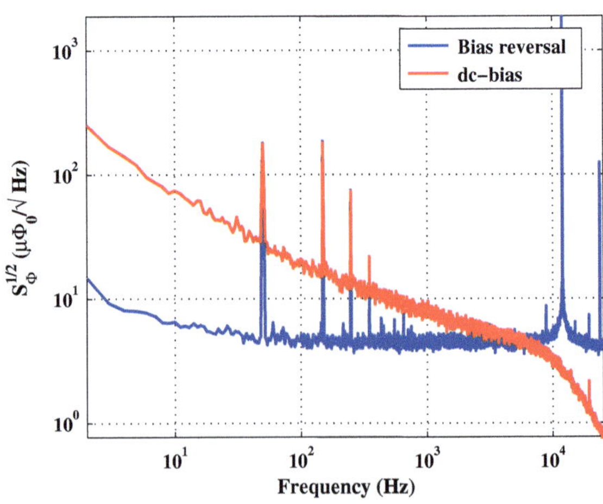

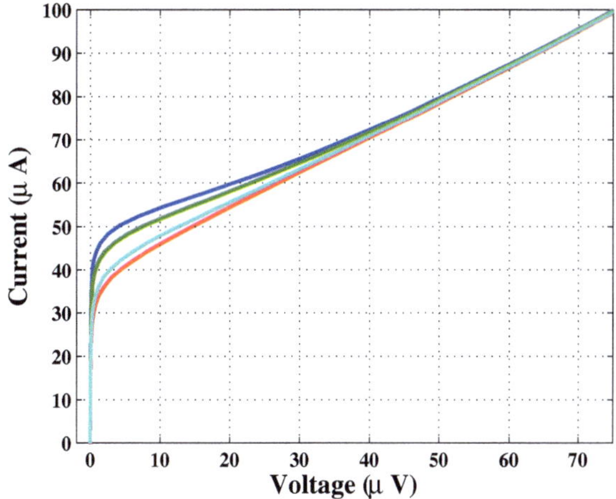

Fig. 2.17 Flux modulated IV curves of GRAD3. The critical current at 77 K of $35\,\mu$A is similar to GRAD2 on the same substrate. The the normal resistance is $1\,\Omega$, giving $I_c R_n = 35\,\mu$V. The maximum voltage modulation is $10\,\mu$V

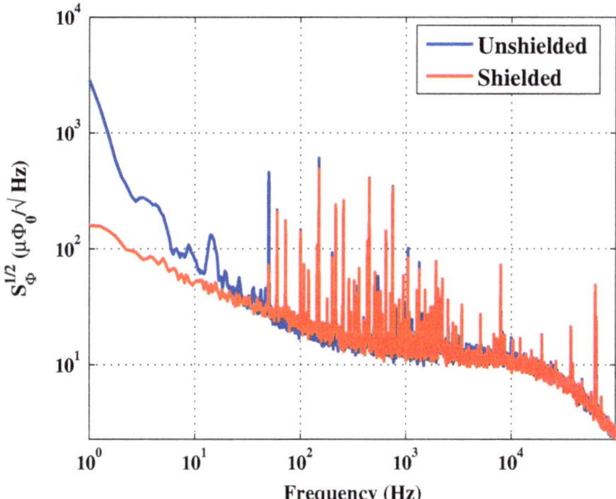

Fig. 2.18 Flux noise of GRAD3 measured with and without shielding with dc bias. The curves nearly overlap indicating that the gradiometer loops are well balanced. The flux noise at 1 kHz is $S_\Phi^{1/2} = 15\,\mu\Phi_0/\sqrt{\text{Hz}}$ with the $1/f$ knee at about 300 Hz

shown in Fig. 2.17. These two SQUIDs were fabricated on the same chip ($10\times10\,\text{mm}^2$ large STO chip with a $24°$ misorientation angle) and their properties were similar. The flux noise of GRAD3 measured in shielded and unshielded environments is shown in Fig. 2.18. Gradiometers 4-5 (GRAD4-5) were fabricated on a STO bicrystal ($24°$ misorientation angle) substrate with a $50\,\text{nm}$ CeO_2 seed-layer. The flux noise of the GRAD4 is shown in Fig. 2.19 measured with and without shielding with dc-bias and bias reversal. The critical current of this SQUID was $220\,\mu$A with $I_c R_n = 176\,\mu$V, and the voltage modulation was $22\,\mu$V at 77 K.

Lastly, GRAD6 was a single gradiometer on a $10\times10\,\text{mm}^2$ large STO chip with a $24°$ misorientation angle. The width of the junctions of the two redundant SQUIDs was $2\,\mu$m and it had flux dams in the design (see Fig. 2.8). The noise of GRAD6L (the left SQUID) is shown in Fig. 2.20 measured in a shielded environment (mu-metal

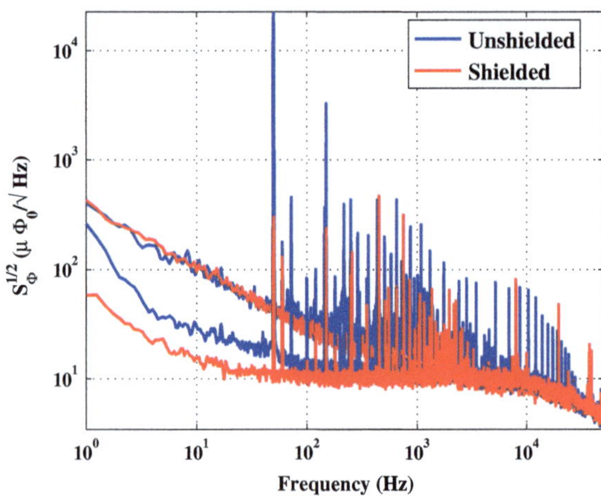

Fig. 2.19 Flux noise of GRAD4 measured with and without shielding. The uppermost curves were measured using dc bias and the remaining two with bias reversal in order to reduce the $1/f$ noise

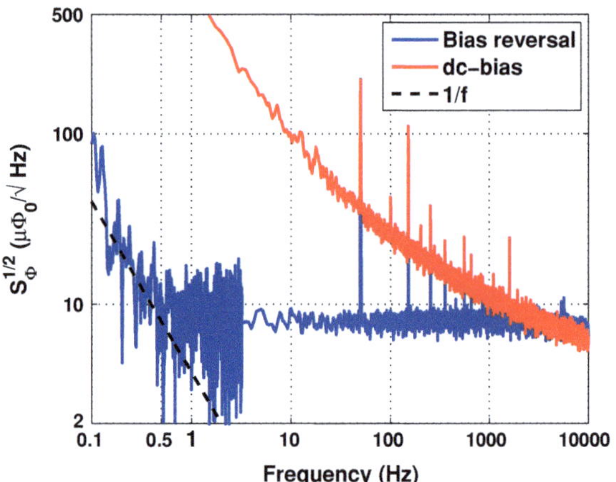

Fig. 2.20 Measured flux noise of GRAD6L with and without bias reversal in a superconducting shield. The $1/f$-knee indicated by the dashed lines is at 0.5 Hz and the white noise level is 7.8 $\mu\Phi_0/\sqrt{\text{Hz}}$

and superconducting shield). The $1/f$-knee is at 0.5 Hz with bias reversal. This is among the best results reported for high-T_c SQUIDs. The low $1/f$-knee with bias reversal shows the absence of trapped fluxes close to the SQUID body, and may be due to the small linewidths close to the SQUID and the presence of flux dams [41].

The most important feature of the magnetometers (MAG1-4, in Table 2.2) used for MEG is that they have low noise at low frequencies (1–20 Hz). The experiments conducted with the magnetometers were performed inside a magnetically shielded room (2 layers of mu-metal and one layer of copper coated aluminum). The SQUIDs were, for MEG recordings, operated with bias-reversal in order to reduce the $1/f$-noise as much as possible. The flux noise of the magnetometer devices was typically 5–8 $\mu\Phi_0/\sqrt{\text{Hz}}$ with the $1/f$-knee below 10 Hz with bias reversal. In order to convert the flux noise to the equivalent magnetic field noise, a magnetic field calibration of the effective area was performed ($A_{eff} = \Phi_0/B$). A 40 cm diameter Helmholtz coil was

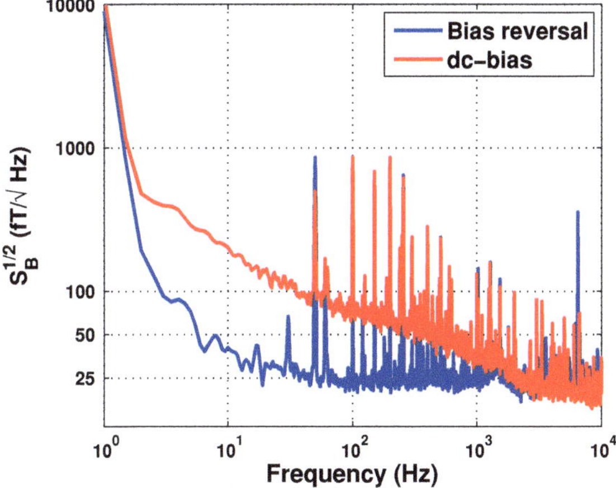

Fig. 2.21 The spectral density of equivalent magnetic field noise of MAG3R measured inside a magnetically shielded room in as the only shielding. The noise above 40 Hz is 25 and 43 fT/$\sqrt{\text{Hz}}$ at 10 Hz with bias reversal

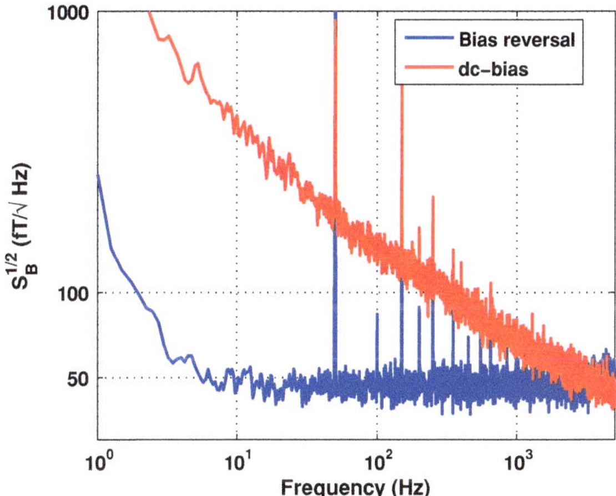

Fig. 2.22 The spectral density of equivalent magnetic field noise of MAG4L measured inside a magnetically shielded room in the MEG setup. The noise at 10 Hz is less than 50 fT/$\sqrt{\text{Hz}}$

placed such that the magnetic field lines were perpendicular to the magnetometer. It is important that the size of the coil is much larger than the sensor area to ensure a homogeneous field. The coil was well calibrated beforehand using a Hall probe magnetometer. The amplitude of the applied field was varied and the output of the SQUID (in FLL-mode) was noted. Finally, the output flux measured by the SQUID magnetometer was plotted against the applied field and with a straight line fit a conversion of 5.3 nT/Φ_0 was obtained. Using the conversion, the magnetic field equivalent noise of two devices is shown in Figs. 2.21 and 2.22 (MAG3R and MAG4L

Table 2.2 A summary of properties of the SQUIDs measured at 77 K

Sample	L (pH)	$2I_c(\mu A)$	β_L	$J_c(A/cm^2)$	$R_n/2(\Omega)$	$I_cR_n(\mu V)$	$\Delta V(\mu V)$	$S_\Phi^{1/2}(\mu\Phi_0/\sqrt{Hz})$
GRAD1	46	125	2.9	10^4	1.3	170	41	4.6
GRAD2	60	35	1.0	4×10^3	1.1	39	11	10
GRAD3	60	35	1.0	4×10^3	1.0	35	10	15
GRAD4	60	220	6.4	1.3×10^4	0.8	176	22	10
GRAD5	60	205	6.0	1.2×10^4	1.0	205	25	9
GRAD6L	60	83	2.4	7×10^3	1.7	141	23	7.8
GRAD6R	60	105	3.0	8.8×10^3	–	–	17	7.5
MAG1L	60	7	0.2	4.3×10^2	8.0	56	30	6.1
MAG1R	60	1.5	0.04	92	14	21	21	80
MAG2L	60	40	1.2	2.5×10^3	2.0	80	30	8.3
MAG2R	60	80	2.4	5×10^3	1.6	128	32	–
MAG3L	60	5^a	0.15	3×10^2	10	50	33	5.7^b
MAG3R	60	5^a	0.15	3×10^2	9.2	46	50	4.7^b
MAG4L	60	80	2.3	4.4×10^3	2.0	160	28	8.2
MAG4R	60	50	1.5	2.8×10^3	2.2	110	32	7.3
MAG5L	60	180	5.2	10^4	–	–	15	12.5
MAG5R	60	80	2.3	4.4×10^3	1.0	80	16	12.5

The notation R and L indicates different SQUIDs on the same chip. The flux noise indicated was at above 1 kHz

[a]Typical operation temperature below 77 K with critical current around $30\,\mu A$

[b]Measured at ~74 K

respectively) measured in the MEG setup. In both devices, the $1/f$-noise was reduced substantially with bias reversal. At 10 Hz the noise was about 50 fT/$\sqrt{Hz}$ or less. In MAG3R, it reached down to 25 fT/$\sqrt{Hz}$ above 40 Hz.

The characteristics of the SQUIDs are summarized in Table 2.2.

References

1. H.K. Onnes, The resistance of pure mercury at helium temperatures. Commun. Phys. Lab. Univ. Leiden, 12(120), (1911).
2. W. Meissner, R. Ochsenfeld, Naturwissenschaften **21**, 787 (1933)
3. L.N. Cooper, Bound electron pairs in a degenerate fermi gas. Phys. Rev. **104**(4), 1189–1190 (1956)
4. J.G. Bednorz, K.A. Müller, Possible high-T_c superconductivity in the Ba-La-Cu-O system. Z. Phys. B **62**(2), 189–193 (1996)
5. M.K. Wu, J.R. Ashburn, C.J. Torng, P.H. Hor, R.L. Meng, L. Gao, Z.J. Huang, Y.Q. Wang, C.W. Chu, Superconductivity at 93 K in a new mixed-phase Y-Ba-Cu-O compound system at ambient pressure. Phys. Rev. Lett. **58**, 908–910 (1987)
6. M. Tinkham, *Introduction to Superconductivity* (Dover publications, Mineola, 1996)
7. T. van Duzer, C.W. Turner, *Superconductive Devices and Electronics* (Prentice Hall PTR, New Jersey, 1999)
8. J. Clarke, A.I. Braginski, The SQUID handbook, vol. 1,(WILEY-VCH, Weinheim, 2006a).
9. J. Clarke, A.I. Braginski, *The SQUID handbook*, vol. 2 (WILEY-VCH, Weinheim, 2006b)

10. B.D. Josephson, Possible new effects in superconductive tunneling. Phys Lett. **1**(7), 251–253 (1962)
11. D.E. McCumber, Effect of ac impedance on dc voltage-current characteristics of supercon-ductor weak-link junctions. J. Appl. Phys. **39**(7), 3113–3118 (1968)
12. W.C. Stewart, Current-voltage characteristics of josephson junctions. Appl. Phys Lett. **12**(8), 277–280 (1968)
13. D. Koelle, R. Kleiner, F. Ludwig, E. Dantsker, J. Clarke, High-transition-temperature super-conducting quantum interference devices. Rev. Mod. Phys. **71**(3), 631–686 (1999)
14. K.K. Likharev, V.K. Semenov, Fluctuation spectrum in superconducting point junctions. JETP Lett. **15**(10), 625–629 (1972)
15. A.N. Vystavkin, V.N. Gubanov, L.S. Kuzmin, K.K. Likharev, V.V. Migulin, V.K. Semenov, S-c-S junctions as nonlinear elements of microwave receiving devices. Phys. Rev. Appl. **9**, 79 (1974)
16. J. Clarke, W.M. Goubau, M.B. Ketchen, Tunnel junction dc SQUID: Fabrication, operation, and performance. J. Low Temp. Phys. **25**, 99–144 (1976)
17. V. Ambegaokar, B.I. Halperin, Voltage due to thermal noise in the dc josephson effect. Phys. Rev. Lett. **22**, 1364–1366 (1969)
18. J. Clarke, R.H. Koch, The impact of high-temperature superconductivity on SQUIDs. Science **242**, 217–223 (1988)
19. R.H. Koch, J. Clarke, W.M. Goubau, J.M. Martinis, C.M. Pegrum, D.J. Van Harlingen, Flicker $(1/f)$ noise in tunnel junction DC Squids. J. Low Temp. Phys. **51**(1–2), 207–224 (1983)
20. J. Clarke, G. Hawkins, Flicker $(1/f)$ noise in Josephson tunnel junctions. Phys. Rev. B **14**(7), 2826–2831 (1976)
21. C.T. Rogers, R.A. Buhrman, Composition of $1/f$ noise in metal-insulator-metal tunnel junc-tions. Phys. Rev. Lett. **53**(13), 1272–1275 (1984)
22. D. Drung, High-T_c and low-t_c dc SQUID electronics. Supercond. Sci. Technol. **16**, 1320–1336 (2003)
23. E. Dantsker, S. Tanaka, J. Clarke, High-T_c superconducting quantum interference devices with slots of holes: Low $1/f$ noise in ambient magnetic fields. Appl. Phys. Lett. **70**(15), 2037–2039 (1997)
24. C.D. Tesche, J. Clarke, dc SQUID: Noise and optimization. J. Low. Temp. Phys. **29**(3–4), 301–331 (1977)
25. K. Enpuku, G. Tokita, T. Maruo, Inductance dependence of noise properties of a high-T_c dc superconducting quantum interference device. J. Appl. Phys. **76**(12), 8180–8185 (1994)
26. K. Enpuku, Y. Shimomura, T. Kisu, Effect of thermal noise on the characteristics of a high T_c dc superconducting quantum interference device. J. Appl. Phys. **73**(11), 7929–7934 (1993)
27. L.P. Lee, J. Longo, V. Vinetskiy, R. Cantor, Low noise $YBa_2Cu_3O_{7-\delta}$ direct-current supercon-ducting quantum interference device magnetometer with direct signal injection. Appl. Phys. Lett. **66**(12), 1539–1541 (1995)
28. P. Magnelind, High-T_c SQUIDs for magnetophysiology, Ph.D. thesis, Chalmers University of Technology, 2006.
29. P.A. Rosenthal, M.R. Beasley, K. Char, M.S. Colclough, G. Zaharchuk, Flux focusing effects in planar thin-film grain-boundary Josephson junctions. Appl. Phys. Lett. **59**(26), 3482–3484 (1991)
30. M.B. Ketchen, J.M. Jaycox, Ultra-low-noise tunnel junction dc SQUID with a tightly coupled planar input coil. Appl. Phys. Lett. **40**(8), 736–738 (1982)
31. M.M. Khapaev, A. Yu, Kidiyarova-Shevchenko, P. Magnelind, M.Y. Kupriyanov. 3D-MLSI: Software package for inductance calculation in multilayer superconducting integrated circuits. IEEE. Trans. Appl. Supercond. **11**(1), 1090–1093 (2001)
32. H. Hilgenkamp, J. Mannhart, Grain boundaries in high-T_c superconductors. Rev. Mod. Phys. **74**, 485–549 (2002)
33. R. Gross, L. Alff, A. Beck, O.M. Froehlich, D. Koelle, A. Marx, Physics and technology of high temperature superconducting josephson junctions. IEEE. Trans. Appl. Supercond. **7**(2), 2929–2935 (1997)

34. R.K. Singh, D. Kumar, Pulsed laser deposition and characterization of high-T_c YBa$_2$Cu$_3$O$_{7-\delta}$. Mater. Sci. Eng. **22**, 113–185 (1997)
35. F. Öisjöen, P. Magnelind, A. Kalaboukhov, D. Winkler, High-T_c SQUID gradiometer system for immunoassays. Supercond. Sci. Technol. 2, 034004 (4pp) (2008).
36. V. Schultze, D. Drung, R. Ijsselsteijn, H-G. Meyer, A high-T_c SQUID gradiometer with integrated homogeneous field compensation. Supercond. Sci. Technol. **17**, S165–S169 (2004)
37. P. Seidel, L. Dörrer, K. Peiselt, F. Schmidl, F. Smidth, C. Steigmeier, Development and investigation of novel single-layer gradiometers using highly balanced gradiometric SQUIDs. Supercond. Sci. Technol. **15**, 150–155 (2002)
38. K. Barthel, D. Koelle, B. Chesca, A.I. Braginski, A. Marx, R. Gross, R. Kleiner, Transfer function and thermal noise of YBa$_2$Cu$_3$O$_{7-\delta}$ direct current superconducting quantum interference devices operated under large thermal fluctuations. Appl. Phys. Lett. **74**(15), 2209–2211 (1999)
39. K. Enpuku, M. Hotta, A. Nakahodo, High-T_c SQUID system for biological immunoassays. Physica C **357–360**(1), 1462–1465 (2001b)
40. F. Ludwig, E. Dantsker, R. Kleiner, D. Koelle, J. Clarke, S. Knappe, D. Drung, H. Koch, N. McN, Alford, T.W. Button. Integrated high-T_c multiloop magnetometer. Appl. Phys. Lett. **66**(11), 1418–1412 (1995)
41. R.H. Koch, J.Z. Sun, V. Foglietta, W.J. Gallagher, Flux dam, a method to reduce extra low frequency noise when a superconducting magnetometer is exposed to a magnetic field. Appl. Phys. Lett. **67**, 709–711 (1995)

Chapter 3
Magnetic Immunoassays

In this chapter the results of the research related to magnetic immunoassays (MIAs) are presented. The relevant background related to the field is introduced before the experimental details and results are discussed.

3.1 Introduction

Immunoassays are used to detect and quantify disease markers in serum, urine or other body fluids. The technique employs the binding reaction between an antibody and its antigen. An early immunoassay technique was radioimmunoassay (RIA) that was described in the early 1960s for insulin by Rosalyn Yalow and Solomon Aaron Berson [1]. Rosalyn Yalow received the 1977 Nobel prize for her invention. In RIA, a radioactive label is used, and in the original study iodine-131 was the label. The usage of radioactive substances gave rise to some concerns regarding safety of personnel and handling of radioactive waste. Furthermore, it was expensive to build special facilities, and the requirements on hardware for this type of system was extensive [2].

The first paper on the todays most commonly used assay, enzyme-linked immunosorbent assay (ELISA), was published in 1971 by Eva Engvall and Peter Perlmann [3]. Almost simultaneously, Anton Schuurs and Bauke van Weemen published a paper on enzyme immunoassay (EIA) [4] that was developed independently of Engvalls and Perlmans work on ELISA. In ELISA/EIA, the radioactive label is replace by an enzyme label. In the late 1970s and early 1980s commercial ELISA/EIA/RIA systems were developed and introduced on the market and since then, they have had a huge impact. For example, the number of articles on ELISA and EIA (based on keywords) was about 5,000 per 5 years in late 1970s and in the late 1990s it was 40,000 and as of 2005, still increasing [2]!

In 1997 Kötitz et al. published their first paper describing an immunoassay system based on magnetic nanoparticles (MNPs) and superconducting quantum

F. Öisjöen, *High-T_c SQUIDs for Biomedical Applications: Immunoassays, Magnetoencephalography, and Ultra-Low Field Magnetic Resonance Imaging*, Springer Theses, DOI: 10.1007/978-3-642-31356-1_3, © Springer-Verlag Berlin Heidelberg 2013

interference devices (SQUIDs) [5]. Since then, several groups have reported on successful immunoassay systems using SQUIDs and MNPs. A common way is to immobilize the MNPs to a substrate [6, 7], to agarose beads [8] or large yeast cells [9]. Strömberg et al. reported on a technique based on rolling circle amplification (RCA) using single-stranded oligonucleotide tagged magnetic nanobeads reaching a sensitivity below 10 pM of RCA coils [10, 11]. Enpuku et al. reached a high sensitivity of 0.5 pg/ml, corresponding to roughly 6×10^6 molecules (or 50 fM), for detection of Interleukin 8 (IL8) using a sandwich-type assay and a SQUID sensor [7]. Eberbeck et al. employed a method where the MNPs bind to large agarose beads (for immobilization) and could quantify 2 nM of streptavidin, or the equivalent to approximately 100 ng/ml [8]. Grossman et al. demonstrated a detection limit of less than 10^6 bacteria cells (*L. monocytogenes*) per ml using freely floating MNPs [12]. High sensitivity to Alzheimer's disease markers in human blood plasma was shown by Yang et al. [13, 14] using MNPs and a high-T_c SQUID.

A number of groups have demonstrated successful bioassay methodologies for detection of biofunctional MNPs using other types of sensor technologies, e.g. an induction coil technique [15–17], frequency mixing [18, 19] and Hall bar sensors [20, 21]. A high iron sensitivity of 0.12 μg/ml was demonstrated by Krause et al. using the frequency mixing technique [18] and an analyte sensitivity of 100 pg/ml was achieved by Nikitin et al. [19]. One technique that does not employ MNPs, apart from ELISA, is based on quartz crystal microbalance (QCM) measurements [22–24]. Using QCM, an analyte sensitivity of 1 pg/ml has been demonstrated [23]. Another example is surface plasmon resonance (SPR) (for a review see [25]) and localized SPR [26–28].

MNPs had been used for over 40 years before Kötitz et al. demonstrated their MNP- and SQUID-based immunoassay system in 1997 [5] and have found their way into a variety of applications [29, 30]. Gilchrist et al. showed in 1957 the feasibility of using MNPs for heating tissue [31], so-called hyperthermia. The MNPs are appropriately functionalized for a target tissue which makes it possible to localize the MNPs close to, e.g. cancerous tissue. An ac magnetic field is applied and a transfer of energy from the MNPs heats the tissue and destroys the cancerous tissue as has been demonstrated several times over the last decades [32, 33]. MNPs are easily manoeuvred due to their magnetic properties which make them suitable for separation and delivery applications. In the context of magnetic separation [34, 35], the MNPs are functionalized such that they capture molecules of interest. The simplest case is to apply a permanent magnet and remove the supernatant liquid leaving only the substance that was targeted. For drug delivery, MNPs are coated with a drug and injected into the blood stream [36–38]. An external magnetic field gathers all the MNPs at a certain location where the drug is to be released. For more details of applications of MNPs, the reader is referred to [29, 30] among many reviews on the topic.

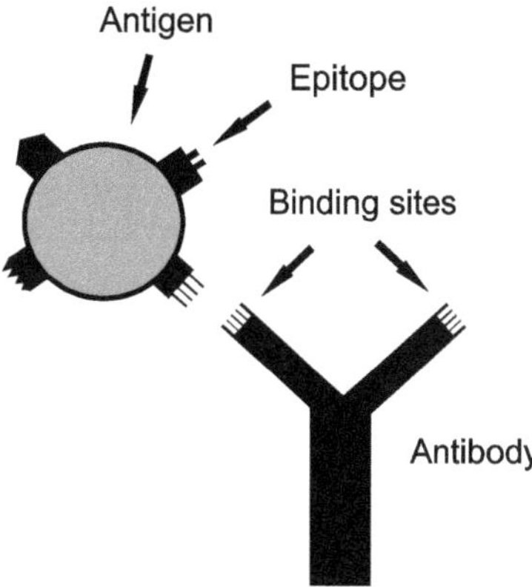

Fig. 3.1 A schematic of a monoclonal antibody and an antigen with four epitopes. The antibody binds with high specificity to one of the epitopes of the antigen

3.1.1 Antibodies and Antigens

Antibodies are produced by the immune system to fight invaders such as foreign bacteria, viruses or other micro-organisms. There are millions of different types of antibodies present in the human body [39] and they are generated by so-called B-cells after they have been activated. B-cells are white blood cells that are produced in the bone marrow. Antibodies on the surface of B-cells are used for recognition of antigens that help the activation of the B-cells (there are other processes involved). The B-cells split into plasma cells that generate the antibodies, and into memory cells with the produced type of antibody on the surfaces as a protection for future infection. The antibodies bind with high specificity to antigens present on the invader. Each antibody has two binding sites and can be of one of two types: monoclonal or polyclonal. A monoclonal antibody can only bind to a certain type of antigen whereas a polyclonal can bind to more than one type. Antigens are present in e.g. infections, cell metabolism, and allergic reactions to mention a few. Antigens have a number of binding sites called epitopes. A schematic of an antibody-antigen complex is shown in Fig. 3.1.

After antibodies bind to antigens, the next step is elimination of the foreign microbe. This is done by a white blood cell commonly known as phagocytes. If a phagocyte encounters an antibody that has bound to an antigen then it captures the antibody-antigen complex and destroys it. For more details on immunology the reader is referred to [39, 40].

3.1.2 ELISA Protocols

ELISA kits are readily available and can be ordered online from several companies. Typical commercial ELISAs reach sensitivities below ng/ml or fM to various types of biomolecules and the cost of an ELISA kit is on the order of 100–1000 USD [41–43].

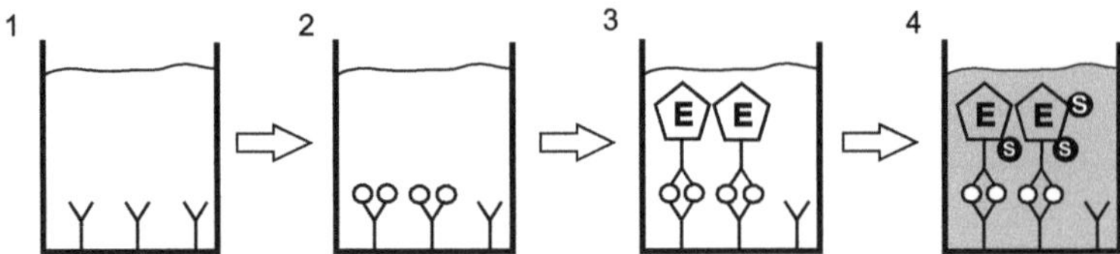

Fig. 3.2 A schematic of a sandwich ELISA. (*1*) The well is coated with the capturing antibodies. After a washing step is performed, the analyte is added and binds to the antibodies in the well (step *2*). After another washing step antibodies bound to an enzyme are added that bind to the other side of the analyte which forms a sandwich structure (*3*). A washing step is performed to remove any residual antibodies and finally (*4*), a chromogenic substrate is added that changes color when it reacts with the enzyme. The intensity of the transmitted light (at the wavelength of the color) is used to determine the amount of analyte present in the sample

There are several protocols for ELISA but the main types are: sandwich [44–46], indirect [46, 47] and competitive [46] ELISAs. Sandwich type ELISAs are the most commonly used and available commercially. What all the protocols have in common though, is that they all rely on several washing steps.

A step by step schematic of a typical sandwich ELISA is shown in Fig. 3.2. The sample one wishes to analyze is added to a well coated with antibodies that the antigens bind to. Antibodies labeled with enzymes are added that bind to the antigen and form the sandwich structure. Finally, a chromogenic substrate is added that will react with the enzyme and make the solution colorful. The intensity of the light determines the amount of antigen that was present in the sample.

In an indirect ELISA the sample of unknown concentration of antigen is added first and will coat the inside of a well. Then detection antibodies are added that bind to the antigen on the surface of the well. These antibodies are usually be pre-labeled with the enzyme. Finally the chromogenic substrate is added and quantification can be made in a similar manner to sandwich ELISA.

Finally, there is a format called competitive ELISA. A well is coated with known antigens and the sample is added simultaneously as the enzyme-labeled antibodies. The antigen bound to the well will compete with the antigen present in the sample to bind with the antibodies. After washing, the enzyme-labeled antibodies that are not bound to the antigen on the wall are removed. The chromogenic substrate is added and in this case, high signal indicates low concentration of antigen present in the sample.

3.2 Magnetic Nanoparticles

In this thesis, MNPs have been used as magnetic labels for antibodies in MIAs. This section introduces the basic terminology of MNPs. These concepts are important in order to understand the results of this research.

3.2.1 General Aspects

The most commonly used magnetic material for MNPs is iron oxide due to its chemical stability and bio-compatibility. Throughout this work, magnetite (Fe_3O_4) and cobalt-ferrite ($CoFe_2O_4$) multi-core MNPs have been used. Other types of MNPs include metals and alloys [48].

MNPs can be either single- or multi-core particles typically embedded in a non-magnetic material such as dextran or starch which enables further functionalization for biomedical applications [49]. Each MNP consists of one single-domain crystal (single-core particle) or several single-domain crystals (multi-core particle). Below a critical size the crystal is in a state of constant magnetization direction, which is defined as a magnetic single-domain crystal. For iron the critical size is about 15 nm [48] and for magnetite the critical size is about 100 nm.

The concept of colloidal stability relies heavily on the quality of the MNP system. A colloidally stable solution consists of a dispersed solid (the particles), that is homogeneously distributed in a liquid phase (e.g. distilled water). Colloidal stability is particularly important for assays based on freely floating MNPs.

3.2.2 Magnetic Anisotropy

An important concept for MNPs is their preference for the magnetic moments of a single-domain crystal to point in a certain direction. The phenomenon is called magnetic anisotropy. In the case of uniaxial symmetry, only two preferred directions of the magnetization are present along a single easy axis. The energy term related to the magnetic anisotropy is given by $E = KV_p \sin^2 \theta$, where K is the anisotropy constant, V_p is the volume of the single domain and θ is the angle between the magnetization and the easy axis [48]. The energy barrier given by KV_p separates the two easy directions of magnetization that are $180°$ apart. There are several effects that cause magnetic anisotropy arising from e.g magneto-crystalline anisotropy, shape anisotropy, stress anisotropy, and surface anisotropy [50].

3.2.3 Size Distribution

MNP systems consist of MNPs with a spread of sizes and the hydrodynamic size distribution of a MNP system can be modeled by a log-normal distribution function given by [51]:

$$f(r_H) = \frac{1}{\sqrt{2\pi} \ln \sigma r_H} \exp[-\frac{(\ln r_H - \ln \bar{r}_H)^2}{2 \ln^2 \sigma}], \qquad (3.1)$$

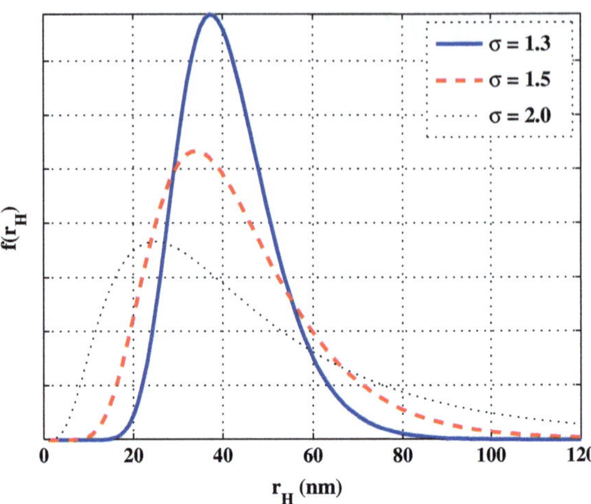

Fig. 3.3 Representative curves of a log-normal size distribution for different standard deviations (σ). The median hydrodynamic radius is kept constant to $\bar{r}_H = 40\,\mathrm{nm}$

where r_H is the hydrodynamic radius of a MNP, $\bar{r}_H$ is the median hydrodynamic radius, and σ denotes the standard deviation of the distribution. Figure 3.3 shows representative log-normal size distributions of varying standard deviations with a fixed median hydrodynamic radius $\bar{r}_H = 40\,\mathrm{nm}$.

3.2.4 Magnetic Relaxation

For MNPs dispersed in liquid, two processes are related to the relaxation of the dynamic magnetic susceptibility refered to as Brownian- and Néel-relaxation. The former is due to stochastic rotation of the particle itself and the latter is related to rotation of the intrinsic magnetic moment of the single-domain crystal over the anisotropy energy barrier.

The Néel relaxation time is given by [52]

$$\tau_N = \tau_0 \exp\left(\frac{K V_p}{k_B T}\right), \tag{3.2}$$

where $\tau_0 = 10^{-9} - 10^{-13}$ [53] is a material-specific characteristic relaxation time that depends on single-domain volume and the temperature, K is the anisotropy constant and V_p is the volume of the single-domain crystal. Correspondingly, the Brownian relaxation time is given by [54]:

$$\tau_B = \frac{4\pi\eta r_H^3}{k_B T}, \tag{3.3}$$

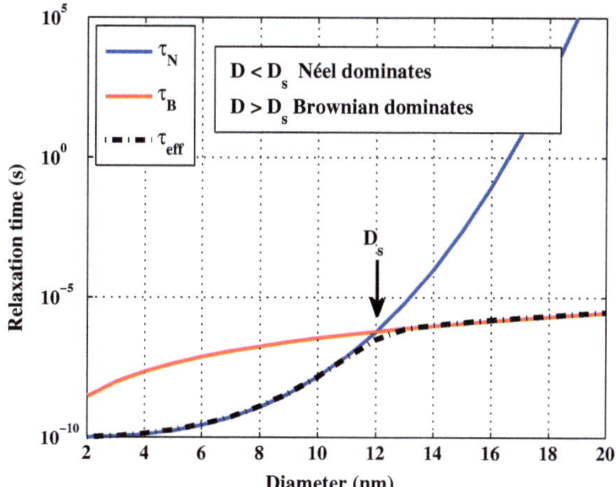

Fig. 3.4 Néel and Brownian relaxation times as a function of single domain diameter at a fixed temperature (300 K) and with an anisotropy constant of 4×10^4 J/m^3 [53], typical for Fe$_3$O$_4$ MNPs. For single domain sizes smaller than the Shliomis diameter (D_s) Néel relaxation is dominating and vice versa for Brownian relaxation

where r_H is the hydrodynamic radius of the MNP, η is the viscosity of the carrier liquid and T is the temperature. Finally, the effective relaxation time is given by [55]

$$\frac{1}{\tau_{eff}} = \frac{1}{\tau_N} + \frac{1}{\tau_B}. \tag{3.4}$$

For a certain single-domain size known as the Shliomis diameter [55] the Néel and Brownian relaxation times are equal. This is illustrated in Fig. 3.4 where D_s indicates the Shliomis diameter given by the diameter where $\tau_N = \tau_B$. Néel relaxation is the dominating relaxation process for single-domain sizes smaller than D_s and larger single-domains will exhibit Brownian relaxation as the essential process.

3.2.5 Superparamagnetism

The energy barrier related to the separation of the easy directions of magnetization that was discussed in the context of magnetic anisotropy is given by KV_p. If the thermal energy, k_BT, is high enough so that the Néel relaxation time (according to Eq. 3.2) is smaller than the measuring time, the magnetization can switch direction faster than what can be observed. This gives the whole magnetic nanoparticle ensemble zero net magnetization in the absence of an applied external magnetic field and no coercivity [56], that defines superparamagnetic particles.

For our type of measurements, we monitor Brownian relaxation, thus, we require thermally blocked MNPs. In a thermally blocked MNP the relaxation time for the single-domains is longer than the Brownian relaxation time. In this case, the total magnetic moment of the MNP rotates with the same rate as the MNP itself. As shown

in Fig. 3.4, the size of the single-domains needs to be larger than about 12 nm in order
to predominantly relax via the Brownian process.

More specifically, in order to distinguish between superparamagnetic particles
and thermally blocked particles, one compares the experimental measurement time,
τ_e with the relaxation time of the internal magnetic moment given by τ_N. A particle
is superparamagnetic if $\tau_N \ll \tau_e$ and thermally blocked if $\tau_N \gg \tau_e$. From these relations it is possible to calculate the blocking temperature, T_B that separates thermally
blocked particles from the superparamagnetic particles. The blocking temperature is
defined by the temperature at which $\tau_N = \tau_e$ for a given volume [56]:

$$T_B = \frac{K V_p}{k_B \ln(\tau_e/\tau_0)}.$$ (3.5)

Similarly, a blocking volume V_B is given by:

$$V_B = \frac{k_B T}{K} \ln(\tau_e/\tau_0)$$ (3.6)

where the temperature is considered to be fixed. Since T_B depends strongly on the
measurement time, τ_e, then different T_B's would be found depending on the measurement technique if the particle volume is fixed. However, if the temperature is kept
constant then the volume of the single domains becomes important. For example, the
blocking radius of magnetite was calculated to be 14 nm for f.c.c cobalt and 12 nm
for iron using a measurement time of 10 s at room temperature [56].

3.3 Experimental Methods

The following section describes the experimental work including measurement techniques, experimental setup, and assay experiments. Two different measurements
techniques were used, ac-susceptometry and magnetorelaxometry, for measurements
of the relaxation dynamics of magnetic nanoparticles and are described in this chapter. Finally, the assay protocols investigated are explained.

3.3.1 Experimental Setup

A photograph of the central part of the experimental setup used for assay experiments
is shown in Fig. 3.5 and a schematic drawing is shown in Fig. 3.6. As indicated, the
SQUID sensor was glued onto a sapphire rod that was in thermal contact with a liquid
nitrogen bath. The cryostat, fabricated by ILK-Dresden, was made of non-magnetic
glass fiber reinforced epoxy. The lid of the cryostat was designed such that a 7 cm
diameter Helmholtz coil, that generated the excitation field H, could surround the

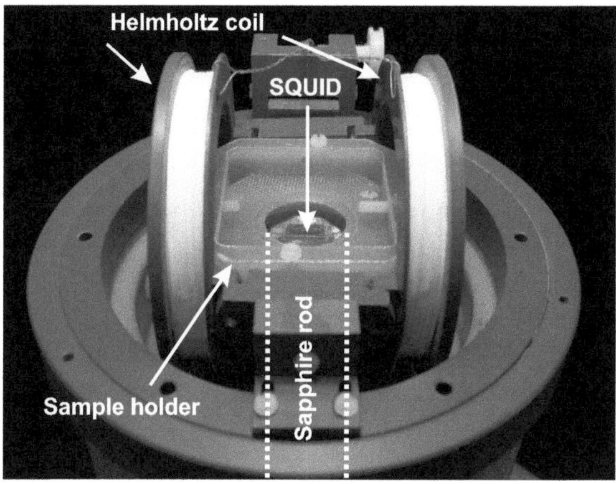

Fig. 3.5 Photograph of the experimental setup. The SQUID is located below the 200 μm thick sapphire window and in the center of the Helmholtz coil. The sample holder frame is located above the SQUID. A glass vial containing a sample of MNPs is put in a hole in a Teflon® piece that is kept in place by the sample holder frame. Adapted from [57]

location of the SQUID and the sample. The sapphire rod, and hence also the SQUID, could be manually aligned close to the 200 μm sapphire window that separated the vacuum-enclosed 77 K SQUID and the room temperature environment. A sensor to sample distance of less than 1 mm was achieved that allowed for a strong coupling between the magnetization M of the sample, that was placed on top of the sapphire window, and the sensor. The Helmholtz coil was manually aligned in order to minimize the coupling between the excitation field, H, and the SQUID. The dc SQUID electronics were the SEL-1 system from Magnicon® with a maximum FLL bandwidth of 6 MHz and a maximum bias reversal frequency of 250 kHz. The electronics used for the particular measurement technique is described separately in the corresponding sections. The ultimate aim for our system is to incorporate a microdroplet handling system. This would allow remote manipulation of tiny volumes of analyte and MNPs. The proof-of-principle microdroplet system that was developed was based on electrowetting-on-dielectric (EWOD), a common microdroplet actuation technique [58]. Briefly, the EWOD chip had both ground and actuation electrodes in the bottom layer. It was coated with a 1.5 μm layer of SU-8 epoxy as the dielectric and finally the top hydrophobic layer was a spin-coated layer of Teflon. For details see Paper III [59] by Schaller et al.

The EWOD technique relies on the polarity of water molecules. As a voltage is applied across two neighboring electrodes on a EWOD chip, the water molecules in the droplet sitting next to the electrodes are attracted by the electric field gradient. The droplet then wets the surface of the chip and, consequently, is transported. For measurements with the SQUID-setup the EWOD chip was placed on the sapphire window with the SQUID close to the window. The EWOD technique can be used for transporting [60], mixing [61], and splitting and merging of droplets [62].

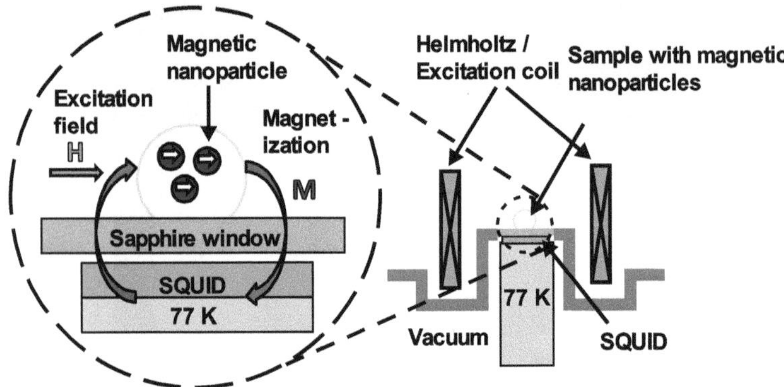

Fig. 3.6   A schematic drawing of the experimental setup. The Helmholtz coil produces the excitation field that magnetizes the sample of MNPs. The small distance between the SQUID and the sample enables strong coupling between the magnetization of MNPs and the SQUID. The figure is not to scale

3.3.2 Measurement Techniques

We used two different measurement techniques: ac-susceptometry (ACS) and magnetorelaxometry (MRX). Regardless of the measurement modality, the aim was to measure changes of the size distribution of MNP systems and ultimately the increase of the median hydrodynamic radius governed by biomolecule binding. Important to remember is that the same experimental setup explained in the previous section was used for both measurement techniques. The following sections describe these in more detail.

AC-Susceptometry

ACS is a frequency domain measurement technique. An external magnetic excitation field magnetizes the MNPs and the stray field from the MNPs is detected with the SQUID. The excitation field applied to the MNP sample is swept over a range of frequencies which yields a frequency dependent response from the MNPs. The magnetic susceptibility, χ, is defined by the relation between an applied magnetic field H and the magnetization, M, of the sample:

$$M = \chi H = (\chi' - i\chi'')H, \tag{3.7}$$

where χ' and χ'' are the real and imaginary parts of the complex magnetic susceptibility, respectively. In order to determine Brownian relaxation times, we consider a model of the dynamic magnetic susceptibility adapted from [16] where it is assumed that the susceptibility of a single particle is equal to the mean value of the susceptibility of the entire ensemble. The dynamic magnetic susceptibility is a complex

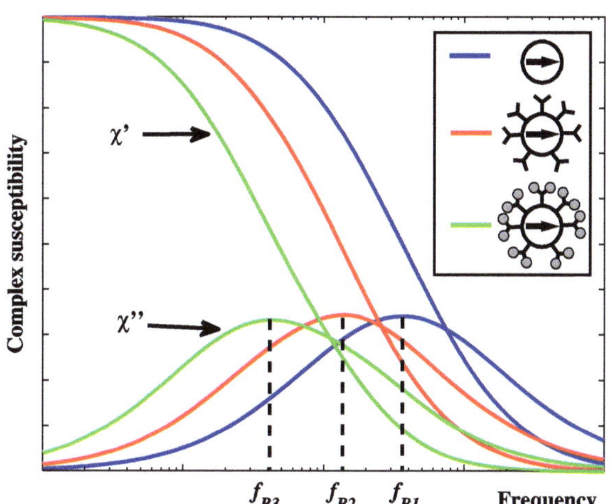

Fig. 3.7 Principle of biomolecule detection using ac-susceptometry. Starting with a reference sample (*blue*) with a Brownian relaxation frequency f_{B1}. Binding of antibodies to the surfaces of the MNPs results in an increase in their hydrodynamic radii that shifts the relaxation dynamics to lower frequencies (*red*), $f_{B2} < f_{B1}$. Further binding of antigens to the antibodies shifts the Brownian relaxation frequency to even lower frequency (*green*), $f_{B3} < f_{B2} < f_{B1}$

function of frequency and is found by using the Debye model [63] assuming the log-normal size distribution given in Eq. 3.1. The model is given by:

$$\chi(\omega) = \langle \chi_{0B} \rangle \int \frac{1}{1 + i\omega\tau_B(r_H)} f(r_H) dr_H + \frac{\chi_{0N}}{1 + (i\omega\tau_N)^\alpha} \tag{3.8}$$

where $\langle \chi_{0B} \rangle$ is the dc susceptibility due to Brownian relaxation, ω the angular frequency, τ_B the Brownian relaxation time, $f(r_H)$ is the distribution of Brownian relaxation times, and r_H is the hydrodynamic radius of the MNPs. The second term comes from the Néel relaxation and is modeled by a Cole-Cole expression where χ_{0N} is the Néel contribution to the static susceptibility, τ_N is the Néel relaxation time, and α is related to the distribution of Néel relaxation times.

The hardware used in ACS measurements include: a Fluke AWG-220 waveform generator for generating the sinusoidal ac magnetic (excitation) field and a Stanford SR-830 lock-in amplifier. The lock-in amplifier is used to measure the real (χ') and imaginary (χ'') parts of the complex susceptibility using the applied signal as the reference. In order to remove the residual coupling between the SQUID and the excitation field a calibration measurement of the background is made. The calibration is subtracted from the measurement of the MNP sample leaving only the magnetic field produced by the MNPs.

Each measurement starts at a low frequency, well below the relaxation frequency of the MNP ensemble. The MNPs can keep up with the applied magnetic field at these low frequencies which means that χ'', that gives the loss of the system, is almost zero. As the frequency approaches the Brownian relaxation frequency, $f_B = 1/2\pi\tau_B$, the magnetization of the MNPs lags behind the applied field and χ'' is no longer non-zero. A peak in χ'' develops at f_B where the energy loss of the system is greatest and χ' has a corresponding roll-off as illustrated in Fig. 3.7.

As biomolecules bind to the surfaces of the MNPs their hydrodynamic size increases according to Eq. 3.3. Consequently, the characteristic Brownian relaxation frequency decreases and appears as a frequency shift between a reference sample (with no biomolecules present) and a sample with biomolecules bound the MNPs as illustrated in Fig. 3.7.

Magnetorelaxometry

MRX is a fast time domain measurement technique. The magnetic field is pulsed in order to magnetize the MNPs and the decay of the magnetization is measured after the pulse is off. The excitation field is generated using a Fluke AWG-220 waveform generator and the signal measured with the SQUID is sampled using a NI-DAQ 6014 A/D converter with a sampling rate of 250 kS/s. A pulse sequence is used such that the feedback of the SQUID electronics is off when the excitation pulse is high. A delay of $\sim$10 µs is introduced in order to allow for stray eddy currents to dissipate and the SQUID feedback electronics to recover.

The time dependence of the decay of the magnetization of a MNP system after applying a pulsed induction field to it can be modeled by [16]:

$$M(t) = M(0) \int \exp\left(\frac{-t}{\tau_B(r_H)}\right) f(r_H) dr_H + M_N \exp\left(-\left(\frac{t}{\tau_N}\right)^{\beta}\right) \quad (3.9)$$

where t is the time measured after the excitation field is switched off and $M(0)$ and M_N is the initial strength of the magnetization due to Brownian and Néel relaxation respectively. The second term in Eq. 3.9 is due to the contribution of Néel relaxation and is modeled with a stretched exponential law with the exponent β related to the distribution of Néel relaxation times.

3.3.3 Assay Experiments

Two types of assays were investigated throughout this work: a cluster type assay and a one-step protocol. The cluster type assay is more complicated and requires more preparations involving incubation and magnetic separation whereas a one-step assay is fast and simple although less sensitive. In our assay experiments we use MNPs functionalized with streptavidin, a molecule with high affinity for biotin [64]. The antibodies used are biotinylated that enables binding of capturing antibodies to the surfaces of the MNPs for further experiments.

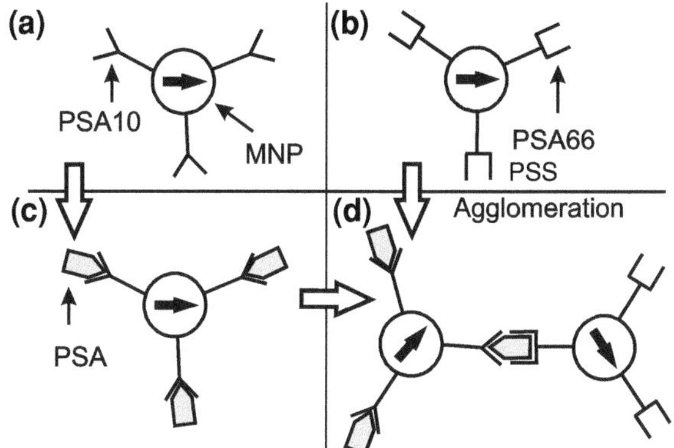

Fig. 3.8 A cartoon of the chemistry of a cluster-type assay (not to scale). Two samples are prepared: one with PSA10 bound to the MNPs and another sample with PSA66 bound to the MNPs (**a** and **b**). The analyte, PSA, is added to the PSA10 sample and bind to the antibodies (**c**). Finally the two are added and forms a sandwich structure (**d**). Adapted from [57]

Cluster-Type Assay

The cluster-type assay relies on the loss of signal caused by relaxation dynamics that are slowed down to fall outside of the measurement window, i.e. the Brownian relaxation time becomes longer than the measurement time. A cartoon of the cluster assay is shown in Fig. 3.8 in a step-by-step manner. We used magnetite (Fe_3O_4) MNPs that were functionalized with streptavidin (Chemicell, FluidMAG®/HS-Streptavidin, particle diameter 100 nm) and suspended in distilled water (DIW). The final MNP mass density of the samples was 2.5 mg/ml. In this immunoassay we targeted prostate-specific antigen (PSA) with a sandwich assay that induced clustering of the MNPs. 100 μl of streptavidin functionalized MNPs were incubated with biotinylated prostate-specific antibodies type 10 (PSA10) for 1 h, see Fig. 3.8a. A similar sample with another prostate specific antibody (PSA66) was prepared (Fig. 3.8b). Sufficient volume and concentration of antibodies were used to coat the MNPs with approximately 20 antibodies/MNP. After incubation for 1 h, a magnetic separation was done by placing a permanent magnet close to the side of the vials containing the MNPs-PSA10(66) and the supernatant liquid was removed using a pipette. A 100 μl volume of 30.2 μg/ml solution of PSA was added to the sample containing MNP-PSA10 and after 1 h of incubation and another magnetic separation step a MNP-PSA10-PSA configuration was obtained as shown in Fig. 3.8c. Finally, MNP-PSA66 was mixed with the MNP-PSA10-PSA complex and clusters of MNPs were formed (Fig. 3.8d).

One-Step Assay

For the one-step assay experiment we used cobalt-ferrite ($CoFe_2O_4$) MNPs functionalized with streptavidin (Chemicell FluidMAG®/HS-Streptavidin, particle diameter 100 nm). This experiment was performed as a calibration of our system in terms of an assay. In each sample we mixed 25 μl of the MNP sample (5 mg/ml mass per fluid

Fig. 3.9 Representative schematic of binding of biotinylated PSA10 to the surface of a MNP functionalized with streptavidin. The figure is not to scale. Adapted from [65]

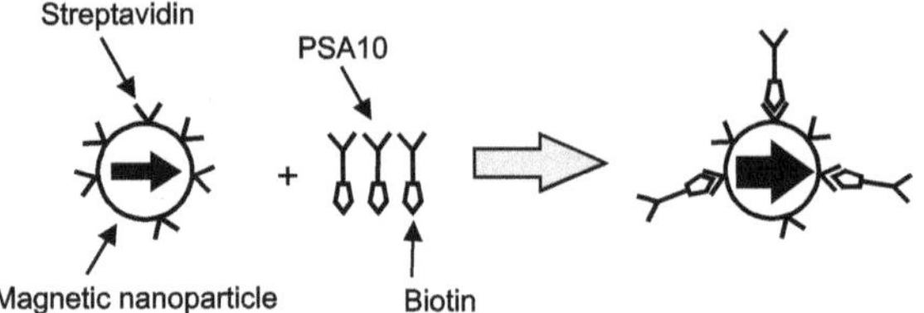

volume) with 25 µl of PSA10 solution of various concentrations. The reference sample was 25 µl of the MNP solution and 25 µl of phosphate buffered saline (PBS). The concentrations of PSA10 were: 60 µg/ml (0.4 µM), 40 µg/ml (0.27 µM), 30 µg/ml (0.2 µM) and 12 µg/ml (0.08 µM). All samples were incubated for 1 h. A cartoon of the biochemistry is shown in Fig. 3.9.

3.4 Results and Discussion

The results included in this section include a system verification, a calibration in terms of iron content sensitivity and MNP sample volume. Moreover, biomolecule sensitivity will be demonstrated for two different assay types described in the previous chapter using ACS and MRX.

3.4.1 System Verification and Calibration

Three separate experiments were made to verify and calibrate our system: MNP content sensitivity measurements, MNP sample volume optimization, and glycerol dilution. Iron content sensitivity is an important figure of merit since it can be used for estimation of the ultimate sensitivity of MIAs. Volume optimization was done in order to maximize the sensitivity to iron for a given SQUID geometry. This is important for the incorporation of the microdroplet handling system. Glycerol dilution was performed in order to verify our system.

MNP Content Sensitivity

The volume of the magnetic fluid (Micromod, nanomag®-D, particle diameter 130 nm, Fe_3O_4) was fixed at 2 µl and the MNPs content was varied. The sample holder for the MNPs was a small hollow cylinder (inner diameter 9 mm, outer diameter 15 mm and height 10 mm) to which plastic foil was mounted at the bottom. A droplet (2 µl) of MNPs was dispensed onto the plastic foil which was replaced after every measurement to avoid contamination. The plastic foil (thickness ~ 10 µm) was not

Fig. 3.10 The SQUID signal as a function of MNP content measured. The sample volume was fixed at $2\,\mu l$ and the concentration was varied. The extrapolated MNP content sensitivity at the noise floor of the SQUID is 1.5 ng at 10 Hz. The SQUID used for this experiment was GRAD5

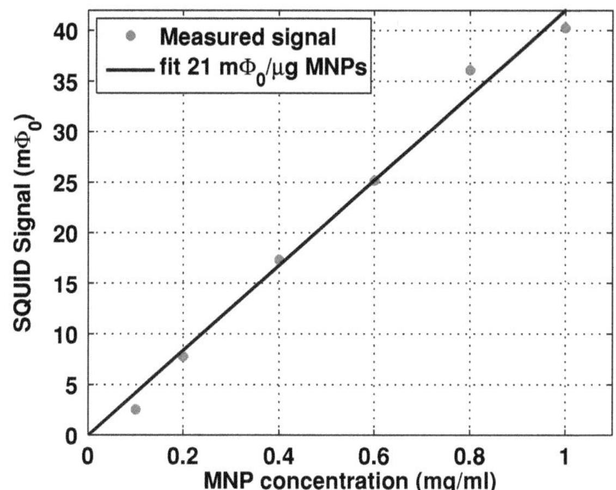

a significant contribution to the separation between the MNPs and the SQUID. Six different concentrations were measured: 1, 0.8, 0.6, 0.4, 0.2 and 0.1 mg/ml of MNP mass per fluid volume. The signal was extracted using the ACS technique at 1, 3.16 and 10 Hz and averaged over the three data points (i.e. at 1, 3.16, and 10 Hz). The data is presented in Fig. 3.10 where a fit to a straight line gives $21\,m\Phi_0/\mu g$ of MNPs. At the noise floor of the SQUID (GRAD5) of about $30\,\mu\Phi_0/\sqrt{Hz}$ at 10 Hz (Fig. 2.18), the extrapolated MNP content sensitivity obtained was 1.5 ng (approximately 2.7×10^6 MNPs) in a $2\,\mu l$ volume. Note that the noise of the SQUID used for this particular measurement was rather high. With a more reasonable noise figure, the extrapolated MNP content sensitivity would reach sub-nanograms.

Microdroplets

The aim of our system was to incorporate a microdroplet handling system. It was therefore important that the signal we measured from droplets was maximized in order to optimize the sample volume with the given SQUID geometry. We performed a calibration measurement in order to experimentally verify that the system was in fact optimized for μl volumes of MNPs. For this experiment, one dilution of MNPs (Micromod, nanomag®-D, particle diameter 130 nm, Fe_3O_4) of 1 mg/ml was prepared and varying volumes of the dilution was measured using the ACS technique. The signal was extracted at low frequencies (1–10 Hz) where all the MNPs are in-phase with the excitation field and the imaginary part is zero. We found that the maximum signal per unit mass of MNPs was for a volume of $2\,\mu l$, as shown in Fig. 3.11. For smaller volumes, the signal was reduced due to the small amount of magnetic material. On the larger volume side, it is a geometrical factor. Some of the MNPs were simply too far away from the SQUID and did not contribute significantly to the signal.

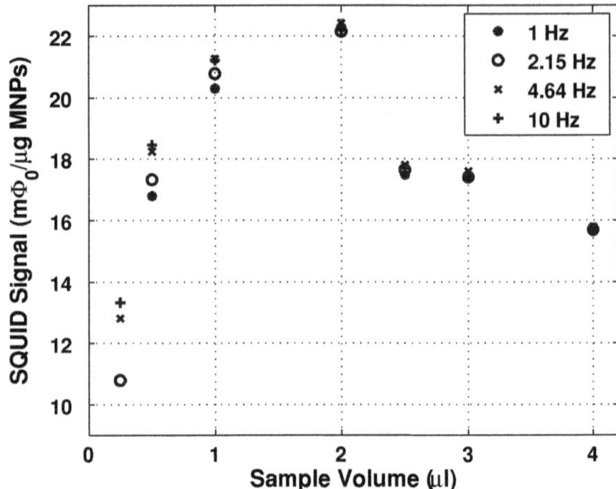

Fig. 3.11 Microdroplet optimization calibration. The highest signal/mass of MNPs was found to be for 2 µl droplets. This data was measured with GRAD5

Glycerol Dilution

In order to verify our system and our measurement techniques a method of controlling the change in relaxation dynamics of a MNP sample was required. The expression for the Brownian relaxation time that was presented in Eq. 3.3 is

$$\tau_B = \frac{4\pi\eta r_H^3}{k_B T},\tag{3.10}$$

and includes the viscosity η. Therefore, by changing the viscosity of the sample the Brownian relaxation time shifts. We used glycerol which has a higher viscosity than DIW to dilute a sample of MNPs and the viscosity of the mixed solution can be found in [66]. In ACS we measure the Brownian relaxation frequency that is inversely proportional to η, i.e. $f_{BRef}/f_{BGlyc} = \eta_{Glyc}/\eta_{Ref}$. In MRX, however, we measure the Brownian relaxation time directly. Altogether, we obtain the following relation between the Brownian relaxation times (frequencies) measured with MRX (ACS) and the viscosities of the two samples [reference (Ref) and glycerol diluted (Glyc)]:

$$\frac{\tau_{BGlyc}}{\tau_{BRef}} = \frac{f_{BRef}}{f_{BGlyc}} = \frac{\eta_{Glyc}}{\eta_{Ref}}.\tag{3.11}$$

The procedure is explained in Paper II and the experimental results measured with GRAD1 are presented in Figs. 3.12 and 3.13 (particles used for this experiment were Chemicell FluidMAG®-D, particle diameter 100 nm).

We measured both samples using ACS and MRX. In ACS, the peaks of the imaginary parts in Fig. 3.12 are, $f_{BRef} = 430$ Hz and $f_{BGlyc} = 160$ Hz and can be used in Eq. 3.11. The median hydrodynamic radius obtained from data fits of the MRX data shown in Fig. 3.13 were used to find the Brownian relaxation times for the two

Fig. 3.12 Frequency domain data from the glycerol dilution experiment. The curve clearly shifts to lower frequencies when the sample is diluted with glycerol to increase the viscosity of the carrier liquid. Adapted from [57]

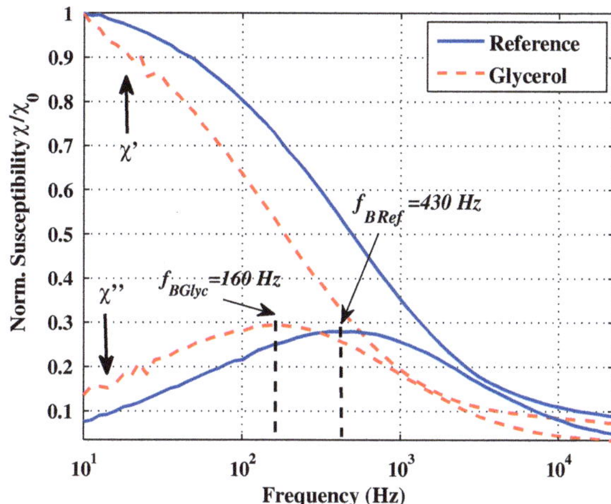

Fig. 3.13 Time domain data from the glycerol dilution experiment. The higher viscosity of the glycerol diluted sample is reflected in the slower relaxation behavior. Adapted from [57]

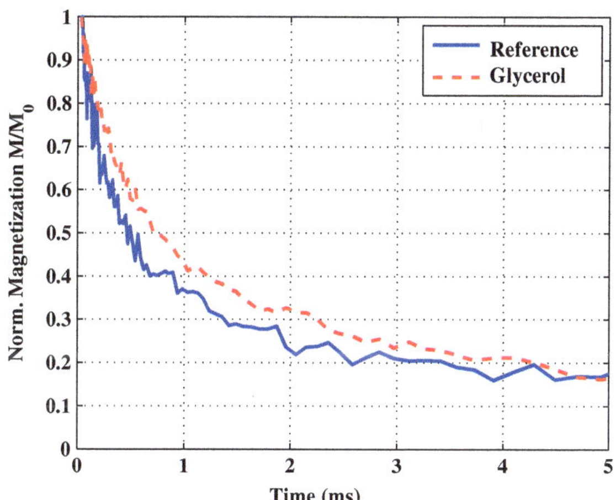

Table 3.1 A summary of the results from the glycerol dilution experiment

Quantity	Value
τ_{BGlyc}/τ_{BRef}	2.5 (MRX)
f_{BRef}/f_{BGlyc}	2.7 (ACS)
η_{Glyc}/η_{Ref}	2.8

samples. We find an excellent agreement and the obtained values are presented in Table 3.1.

3.4.2 Assay Experiments

In this thesis two different assays are described in this chapter that are different in regards to both the detection technique and the biochemical methodology. Cluster type assays rely on the loss of signal whereas the one-step type measures changes of signal. In general, more involved chemistry enables a higher specific sensitivity. For example, the cluster-type experiment performed required two separate samples of MNP-antibody binding to be prepared and two separate antibody-antigen bindings to occur. The one-step assay would require a single sample to be prepared to which the antigen is added. This can be done with fewer washing steps which reduces the risk of residual contamination and loss of reagent. The results from the two assays are presented in the following sections.

Cluster-Type Assay

The cluster-type assay was the more complicated in which two types of antibodies were bound to two separate samples of MNPs. The PSA analyte was added to one of the samples and finally the two were mixed and clustering occured. Data obtained with ACS and MRX are presented in Figs. 3.14 and 3.15 respectively. The reference trace in the ACS data shows a peak, although somewhat broad, at around 250 Hz. The second trace does not have a peak within the measurement window indicating clustering of the MNPs induced by antibody-antigen binding. In Fig. 3.15 data obtained by measuring the same samples using MRX is presented. It is clear that the relaxation is substantially slowed down as a result of the cluster formation. The sensitivity extrapolated to the noise limit of the SQUID used in this experiment (GRAD1) is 18 ng/ml corresponding to roughly 4×10^{10} PSA molecules in a 100 μl sample. This estimation was made using the loss of signal in the imaginary part of the ACS data in Fig. 3.14 at 180 Hz. This experiment is discussed in Paper II.

One-Step Assay

The second type of protocol that was used was a one-step assay. This experiment was also used as a verification and validation of our system in terms of analyte sensitivity. Different amounts of biotinylated antibodies were mixed with streptavidin functionalized MNPs. Samples were measured with our two techniques (using GRAD1) but also at Imego AB with their commercially available DynoMag® based on an induction coil method [16, 17]. These results are presented in Paper IV but I will briefly summarize them here. In Figs. 3.16 and 3.17 representative data of ACS and MRX measurements respectively are presented. A clear shift in frequency as a function of concentration of PSA10 is observed in the frequency domain traces. This was a result of an increased amount of antibodies that bound to the surfaces of the MNPs leading to an increase of the mean hydrodynamic radius of the system. The

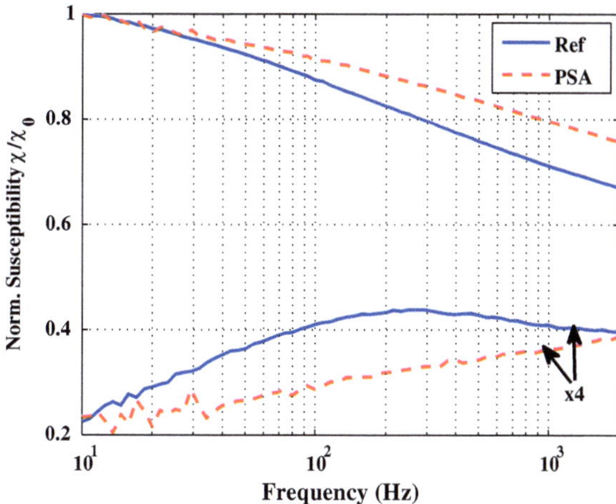

Fig. 3.14 Cluster experiment sample and reference sample measured with ACS. The peak of the imaginary part is visible in the reference trace. In the PSA trace, it is shifted towards lower frequencies outside the measurement window. The imaginary parts are magnified 4 times for clarity. Adapted from [57]

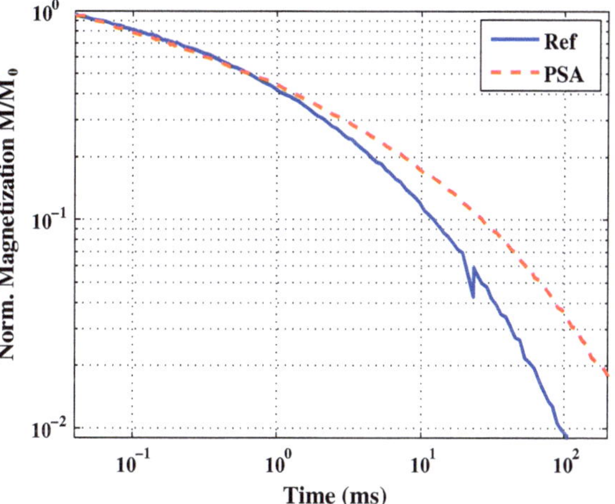

Fig. 3.15 Normalized magnetization measured with MRX on cluster assay experiment samples. As indicated in the ACS data, relaxation times are clearly different in the two samples. Adapted from [57]

same trend can be seen in the time domain data where the relaxation time constant increased with increasing concentration of antibodies.

The data was fit to the models described in Sect. 3.3.2 and information about the size distribution was extracted. The mean hydrodynamic radius as a function of concentration of PSA 10 is shown in Fig. 3.18 where data obtained with DynoMag® is also included. From this figure we concluded that approximately $10\,\mu\text{g/ml}$ is the lower limit for detection corresponding to about 10^{12} molecules. These measurements were performed with GRAD1 presented in Paper I and in Fig. 2.16.

The MNP sensitivity of 1.5 ng that was extracted from the MNP content sensitivity calibration (see Sect. 3.4.1) corresponds to approximately 10^6 MNPs which is about 100,000 times less than the number of MNPs used in the one-step assay. In addition, a rather large volume of MNPs was used. If the sample volume is reduced to $2\,\mu\text{l}$ (i.e.

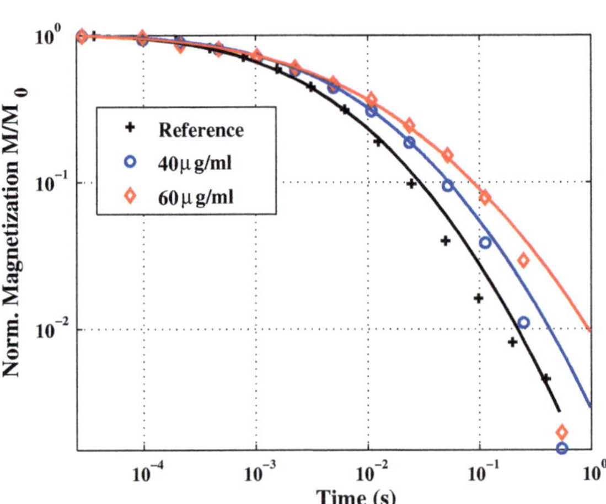

Fig. 3.16 *Curves* representing ACS data of samples containing different amounts of PSA10 in the one-step type assay. The *lines* are data fits to the corresponding *curves* using the model for ACS discussed in Chap. 4. Adapted from [65]

Fig. 3.17 MRX data of three different concentrations of PSA10 in the one-step type assay. The *lines* are data fits to the corresponding *curves* using the model for MRX discussed in Chap. 4. Adapted from [65]

1 µl of MNPs and 1 µl of PSA10), then the number of PSA10 molecules (at the limit, 10 µg/ml) scales to 4×10^{10} if the final MNP concentration is fixed at 5 mg/ml after PSA10 and MNPs have been mixed (the same as in this experiment). The number of MNPs in a 2 µl droplet with a concentration of 5 mg/ml is roughly 10^{10}. At the detection limit of MNP content (1.5 ng, 10^6 MNPs, Sect. 3.4.1) we can extrapolate a sensitivity to PSA molecules, assuming it scales with the MNP content, of 4×10^6 corresponding to 3.5 pM or 500 pg/ml in a 1 µl PSA10 sample volume.

Suggested Binding Rate Measurements

One possibility of our immunoassay system that was not implemented was to utilize the fast MRX-technique for studies of binding rates [8, 12]. For example,

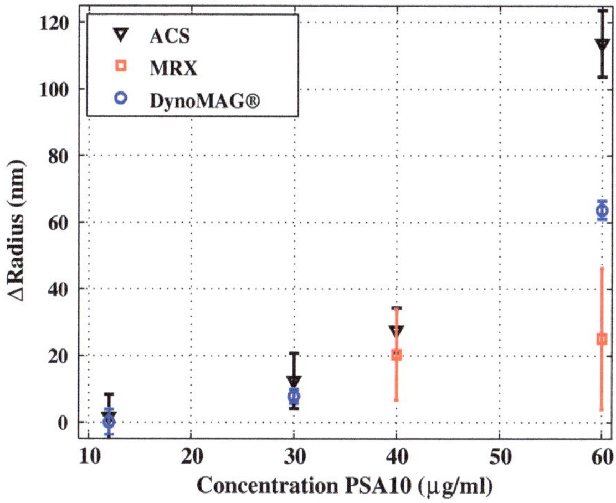

Fig. 3.18 The change in median hydrodynamic radius, as extracted from data fits, as a function of concentration of PSA10 added to separate samples of functionalized MNPs measured with ACS, MRX, and Imego AB's DynoMag®. The data points coincide well except for the high concentration data point. This can be related to differences in measurement setup and techniques. Adapted from [65]

Fig. 3.19 Measurement sequence for tracking of binding events using the ACS and MRX techniques

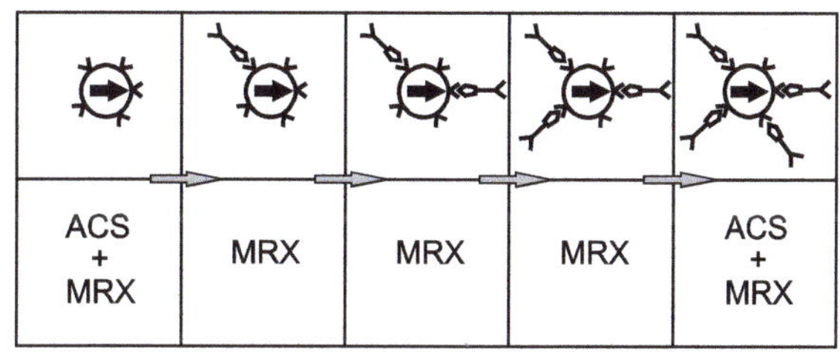

one could study how drug-protein interactions affect the binding activity [67] or the binding kinetics of antibody-antigen binding [68, 69]. A possible measurement sequence is described in Fig. 3.19. First, a reference trace is measured with ACS in the frequency domain of the reference sample. As the analyte is added, MRX data is taken continuously until the reaction appears to have completed. Finally, a final ACS measurement is performed. With this technique, the binding can be tracked in real-time.

3.4.3 Aging of Functionalized MNPs

An observation made during measurements on subsequent days was a significant agglomeration behavior of the functionalized MNPs. This issue was addressed in Paper V where the need for stable functionalized MNPs for immunoassays based on liquid phase homogeneous assays is discussed. The main result is shown in Fig. 3.20 where time-dependent agglomeration was observed. Agglomeration of MNPs is a known problem that induces a number of issues for MNP-based bioassays. It is particularly important for a one-step assay where changes in hydrodynamic radius of freely-floating MNPs is measured. Firstly, the relative change in median hydro-

Fig. 3.20 The change in median hydrodynamic radius of two batches of functionalized MNPs as a function of time where r_0 refers to the radius on day 0. The particles have a median hydrodynamic radius of 50 nm before functionalization. The inset shows the change in the standard deviation of the distribution. Reprinted with permission from [70]

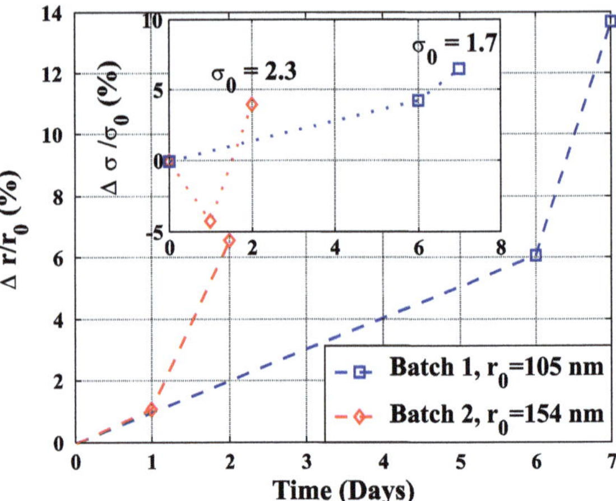

dynamic radius due to biomolecule binding to the surfaces of the MNPs will be smaller if the MNPs have agglomerated which affects the apparent sensitivity of the assay. Secondly, the surface area available for binding of biomolecules decreases which reduces the specific sensitivity further. Thirdly, time dependent changes in hydrodynamic radii of the MNPs could introduce artifacts in the measurements since detection of biomolecules relies on changes of median hydrodynamic radius. Finally, clustering will shift the relaxation dynamics to lower frequencies that increases the relaxation time and therefore also the measurement time substantially.

References

1. R.S. Yalow, S.A. Berson, Immunoassay of endogenous plasma insulin in man. J. Clin. Invest. **39**(7), 1157–1175 (1960)
2. R.M. Lequin, Enzyme immunoassay (EIA)/Enzyme-linked immunosorbent assay (ELISA). Clin. Chem. **51**(12), 2415–2418 (2005)
3. E. Engvall, P. Perlmann, Enzyme-linked immunosorbent assay (ELISA). Quant. assay immunoglobul. G. Immunochem. **8**(9), 871–874 (1971)
4. B.K. van Weeman, A.H.W.M. Schuurs, Immunoassay using antigen enzyme conjugates. FEBS Lett. **15**(3), 232–236 (1971)
5. R. Kötitz, H. Matz, L. Trahms, H. Koch, SQUID based remanence measurements for immunoassays. IEEE Trans. Appl. Supercond. **7**(2), 3678–3681 (1997)
6. K. Enpuku, T. Minotani, M. Hotta, A. Nakahodo, Application of high T_c SQUID magnetometer to biological immunoassays. IEEE Trans. Appl. Supercond. **11**(1), 661–664 (2001)
7. K. Enpuku, A. Ohba, K. Inoue, T.Q. Yang, Application of HTS SQUIDs to biological immunoassays. Phys. C **412–414**, 1473–1479 (2004)
8. D. Eberbeck, C. Bergemann, F. Wiekhorst, U. Steinhoff, L. Trahms, Quantification of specific bindings of biomolecules by magnetorelaxometry. J. Nanobiotechnol. **6**(4), (2008).
9. D. Eberbeck, C. Bergemann, S. Hartwig, U. Steinhoff, L. Trahms, Binding kinetics of magnetic nanoparticles on latex beads and yeast cells studied by magnetorelaxometry. J. Magn. Magn. Mater. **289**, 435–438 (2005)

10. M. Strömberg, J. Göransson, K. Gunnarsson, M. Nilsson, P. Svedlindh, M. Strømme, Sensitive molecular diagnastics using volume-amplified magnetic nanobeads. Nanoletters 8(3), 816–821 (2008).
11. T.Z.G. de la Torre, A. Mezger, D. Herthnek, C. Johansson, P. Svedlindh, M. Nilsson, M. Strømme, Detection of rolling circle amplified DNA molecules using probe-tagged magnetic nanobeads in a portable AC susceptometer. Biosens. Bioelectron. 29, 195–199 (2011)
12. H.L. Grossman, W.R. Myers, V.J. Vreeland, R. Bruehl, M.D. Alper, C.R. Bertozzi, J. Clarke, Detection of bacteria in suspension by using a superconducting quantum interference device. Proc. Natl. Acad. Sci. USA 101(1), 129–134 (2004)
13. C. Yang, S. Yang, J. Chieh, H. Horng, C. Hong, H. Yang, K.H. Chen, B.Y. Shih, T. Chen, M. Chiu, Biofunctionalized magnetic nanoparticles for specifically detecting biomarkers of Alzheimer's disease in vitro. ACS Chem. Neurosci. 2, 500–505 (2011)
14. S.Y. Yang, J.J. Chieh, W.C. Wang, C.Y. Yu, C.B. Lan, J.H. Chen, H.E. Horng, C.-Y. Hong, H.C. Yang, W. Huang, Ultra-highly sensitive and wash-free bio-detection of H5N1 virus by immunomagnetic reduction assays. J. Virological. Methods 153, 250–252 (2008)
15. A.P. Astalan, F. Ahrentorp, C. Johansson, K. Larsson, A. Krozer, Biomolecular reactions studied using changes in Brownian rotation dynamics of magnetic particles. Biosens. Bioelectron. 19, 945–951 (2004)
16. A.P. Astalan, C. Jonasson, K. Petersson, J. Blomgren, D. Ilver, C. Johansson, Magnetic response of thermally blocked magnetic nanoparticles in a pulsed magnetic field. J. Magn. Magn. Mater. 311, 166–170 (2007)
17. A.P. Astalan, J. Blomgren, K. Petersson, C. Jonasson, D. Ilver, C. Johansson, A. Krozer, Brownian relaxation measurements in the time domain of thermally blocked magnetic nanoparticles. Eurosensors XX proceedings, T3C–P6, 2006.
18. H.-J. Krause, N. Wolters, Y. Zhang, A. Offenhäusser, P. Miethe, M. Meyer, M. Hartmann, M. Keusgen, Magnetic particle detection by frequency mixing for immunoassay applications. J. Magn. Magn. Mater. 311, 436–444 (2007)
19. P.I. Nikitin, P.M. Vetoshko, T.I. Ksenevich, New type of biosensor based on magnetic nanoparticle detection. J. Magn. Magn. Mater. 311, 445–449 (2007)
20. L. Ejsing, M.F. Hansen, K. Menon, H.A. Ferreira, D.L. Graham, P.P. Freitas, Planar Hall effect sensor for magnetic micro and nanobead detection. Appl. Phys. Lett. 84(23), 4729–4731 (2004)
21. P.-A. Besse, G. Boero, M. Demierre, V. Pott, R. Popovic, Detection of a single magnetic microbead using a miniaturized silicon hall sensor. Appl. Phys. Lett. 80(22), 4199–4201 (2002)
22. S.-H. Lee, D.D. Stubbs, J. Cairney, W.D. Hunt, Rapid detection of bacterial spores using a quartz crystal microbalance (QCM) immunoassay. IEEE Sens. J. 5(4), 737–743 (2005)
23. K. Yokohama, K. Ikebukuro, E. Tamiya, I. Karube, N. Ichiki, Y. Arikawa, Highly sensitive quartz immunosensors for multisample detection of herbicides. Anal. Chim. Acta 304, 139–145 (1995)
24. Y.-C. Liu, C.-M. Wang, K.-P. Hsiung, Comparison of different protein immobilization methods on quartz crystal microbalance surface in flow injection immunoassay. Anal. Biochem. 299, 130–135 (2001)
25. J. Homola, Present and future surface plasmon resonance biosensors. Anal. Bioanal. Chem. 377, 528–539 (1998)
26. E.M. Larsson, J. Alegret, M. Käll, D.S. Sutherland, Sensing characteristics of NIR localized surface plasmon resonances in gold nanorings for applications as ultrasensitive biosensors. Nano Lett. 5, 1256–1263 (2007)
27. S. Chen, M Svedendahl, M. Käll, L Gunnarsson, A. Dimitriev, Ultrahigh sensitivity made simple: nanoplasmonic label-free biosensing with an extremely low limit-of-detection for bacterial and cancer diagnostics. Nanotechnology 20, 434015 (2009)
28. M.P. Jonsson, A.B. Dahlin, P. Jönsson, F. Höök, Nanoplasmonic biosensing with focus on short-range ordered nanoholes in thin metal films. Biointerphases 3, FD30-FD40 (2009).

29. Q.A. Pankhurst, J. Connolly, S.K. Jones, J. Dobson, Applications of magnetic nanoparticles in biomedicine. J. Phys. D: Appl. Phys. **36**, R167–R181 (2003)
30. A.K. Gupta, M. Gupta, Synthesis and surface engineering of iron oxide nanoparticles for biomedical applications. Biomaterials **26**(18), 3995–4021 (2005)
31. R.K. Gilchrist, R. Medal, W.D. Shorey, R.C. Hanselman, J.C. Parrott, C.B. Taylor, Selective inductive heating of lymph nodes. Ann. Surg. **146**(4), 596–606 (1997)
32. N.F. Borrelli, A.A. Luderer, J.N. Panzarino, Hysteresis heating for the treatment of tumours. Phys. Med. Biol. **29**(5), 487–494 (1984)
33. N.A. Brusentsov, L.V. Nikitin, T.N. Brusentsova, A.A. Kuznetsov, F.S. Bayburtskiy, L.I. Shumakov, N.Y. Jurchenko, Magnetic fluid hyperthermia of the mouse experimental tumor. J. Magn. Magn. Mater. **252**, 378–380 (2002)
34. H. Gu, K. Xu, C. Xu, B. Xu, Biofunctional magnetic nanoparticles for protein separation and pathogen detection. Chem. Commun. **37**, 941–949 (2006)
35. S. Morisada, N. Miyata, K. Iwahori, Immunomagnetic separation of scum-forming bacteria using polyclonal antibody that recognizes mycolic acids. J. Microbio. Meth. **51**, 141–148 (2002)
36. A.K. Gupta, A.S.G. Curtis, Surface modified superparamagnetic nanoparticles for drug delivery: Interaction studies with human fibroblasts in culture. J Mat Sci: Mat. Med. **15**, 493–496 (2004)
37. M. Arruebo, R. Fernández-Pacheco, M.R. Ibarra, J. Santamaría, Magnetic nanoparticles for drug delivery. Nano Today **2**(3), 22–32 (2007)
38. C. Alexiou, R.J. Klein, W. Arnold, F.G. Parak, P. Hulin, C. Bergemann, W. Erhardt, S. Wagenpfeil, A.S. Lübbe, Locoregional cancer treatment with magnetic drug targeting. Cancer Res. **60**, 6641–6648 (2000)
39. H. Brändén, J. Andersson, *Grundläggande immunologi* (Studentlitteratur, Lund, 1998)
40. P. Lydyard, A. Whelan, M.W. Fanger, *Immunology* (Garland Science/BIOS Scientific publisher, London, 2004)
41. http://www.genwaybio.com/all_elisa_kits.php
42. http://www.promokine.info/products/antibodies-elisa-kits/elisa-kits/
43. http://www.cellsignal.com/ddt/elisa_line.html
44. S. Mukherjee, A. Casadevall, Sensitivity of sandwich enzyme-linked immunosorbent assay for Cryptococcus neoformans polysaccharide antigen is dependent on the isotypes of the capture and detection antibodies. J. Clin. Microbiol. **33**(3), 765–768 (1995)
45. H. Ueda, K. Tsumoto, K. Kubota, E. Suzuki, T. Nagamune, H. Nishimura, P.A. Schueler, G. Winter, I. Kumagai, W.C. Mahoney, Open sandwich ELISA: A novel immunoassay based on the interchain interaction of antibody variable region. Nat. Biotechnol. **14**, 1714–1718 (1996)
46. J. MacCarthy, *Detecting Pathogens in Food* (Woodhead Publishing Limited, Cambridge, 2003)
47. C. Björkman, O.J.M. Holmdal, A. Uggla, An indirect enzyme-linked immunoassay (ELISA) for demonstration of antibodies to neospora caninum in serum and milk of cattle. Vet. Parasitol. **68**, 251–260 (1997)
48. A.-H. Lu, E.L. Salabas, F. Schuth, Magnetic nanoparticles: synthesis, protection, functionalization, and application. Angew. Chem. Int. Ed. **46**, 1222–1244 (2007)
49. V. Schaller, Magnetic multicore nanoparticles: properties and applications. Ph.D. Thesis, Chalmers University of Technology, 2010.
50. S. Chikazumi, *Physics of Ferromagnetism* (Oxford University Press, Oxford, 1997)
51. C.G. Granqvist, R.A. Buhrman, Ultrafine metal particles. J. Appl. Phys. **47**(5), 2200–2219 (1976)
52. L. Néel, Compt. Rend. **228**(8), 664–666 (1949)
53. C. Johansson, Magnetic studies of magnetic liquids. Ph.D. Thesis, Department of physics, Chalmers University of Technology and University of Göteborg, 1993.
54. W.F. Brown, Thermal fluctuations of a single-domain particle. Phys. Rev. **130**(5), 1677–1686 (1963)

55. M.I. Shliomis, Magnetic fluids. Sov. Phys. Uspekhi **17**(2), 153–169 (1974)
56. C.P. Bean, J.D. Livingston, Superparamagnetism. J. Appl. Phys. **30**(4), 120S–129S (1959)
57. F. Öisjöen, J.F. Schneiderman, M. Zaborowska, S. Karthikeyan, P. Magnelind, A. Kalabukhov, K. Petersson, A.P. Astalan, C. Johansson, D. Winkler, Fast and sensitive measurement of specific antigen-antibody binding reactions with magnetic nanoparticles and HTS SQUID. IEEE Trans. Appl. Supercond. **19**, 848–852 (2009)
58. F. Mugele, J.-C. Baret, Electrowetting: from basics to applications. J. Phys. Condens. Matter. **17**, R705–R774 (2005)
59. V. Schaller, A. Sanz-Velasco, A. Kalabukhov, J.F. Schneiderman, F. Öisjöen, A. Jesorka, A.P. Astalan, A. Krozer, C. Rusu, P. Enoksson, D. Winkler, Towards an electrowetting-based digital microfluidic platform for magnetic immunoassays. Lab on chip **9**, 3433–3436 (2009)
60. M.G. Pollack, R.B. Fair, A.D. Shenderov, Electrowetting-based actuation of liquid droplets for microfluidic applications. Appl. Phys. Lett. **77**, 1725–1726 (2000)
61. P. Paik, V.K. Pamula, R.B. Fair, Rapid droplet mixers for digital microfluidic systems. Lab Chip **3**, 253–259 (2003)
62. S.K. Cho, H. Moon, C.-J. Kim, Creating, transporting, cutting, and merging liquid droplets by electrowetting-based actuation for digital microfluidic circuits. J. Microelectromech. Syst. **12**, 70–80 (2003)
63. P. Debye, *Polar Molecules* (The chemical catalog company, Haverhil, 1929)
64. E.P. Diamandis, T.K. Christopoulos, The biotin-(strept)avidin system: Principles and applications in biotechnology. Clin. Chem. **37**(5), 625–636 (1991)
65. F. Öisjöen, J.F. Schneiderman, A.P. Astalan, A. Kalabukhov, C. Johansson, D. Winkler, A new approach for bioassays based on frequency- and time-domain measurements of magnetic nanoparticles. Biosens. Bioelectron. **25**, 1008–1013 (2010)
66. R.C. Weast, M.J. Astle, CRC Handbook of Chemistry and Physics, 59th edn. (CRC Press Inc., Boca Raton, 1978–1979).
67. T.C. Kwong, Free drug measurements: methodology and clinical significance. Clin. Chim. Acta **151**, 193–216 (1986)
68. M.B. Medina, L. Van Houten, P.H. Cooke, S.I. Tu, analysis of antibody binding interactions with immobilized E. coli O157:H7 cells using the BIA core. Biotechnol. Tech. **11**(3), 173–176 (1997)
69. S.S. Pathak, H.F.J. Savelkoul, Biosensors in immunology: the story so far. Immunol. Today **18**(10), 464–467 (1997)
70. F. Öisjöen, J.F. Schneiderman, A.P. Astalan, A. Kalabukhov, C. Johansson, D. Winkler, The need for stable, mono-dispersed, and biofunctional magnetic nanoparticles for one-step magnetic immunoassays. J. Phys. Conf. Ser. 200(122006), (2010b). doi:10.1088/1742-6596/200/12/122006

Chapter 4
Magnetoencephalography

This chapter describes the experiments related to magnetoencephalography with high-T_c SQUIDs. Necessary background on basic brain physiology and neurons will be introduced before the experimental details and results are discussed. For a comprehensive review of MEG, both technical and methodological, see [1].

4.1 Introduction

A neuron is the most fundamental cell of the brain and generates electrical activity in order to communicate with other neurons or parts of the body. Magnetoencephalography (MEG) is the measurement of the magnetic fields generated by neural activity in the brain. The corresponding technique for the electric field is electroencephalography (EEG). EEG has a long history and the first findings were reported by Caton in 1875 when he measured electrical activity in the brains of rabbits and monkeys [2]. Berger was the first to record a human EEG (he also gave the technique its name) in 1929 [3]. Since these remarkable achievements, EEG has evolved to become an important tool and is widely used for both scientific and clinical purposes. In the clinic it is particularly important for characterization of epileptic seizures [4].

Measurements of the magnetic field generated by neural currents with sufficient signal to noise ratios require extremely sensitive detectors. The first MEG recordings were made with copper coils by Cohen in 1968 [5]. The induction readout technique, however, meant the SNR was very low. A few years later the SQUID was invented [6], and in 1972 Cohen published the first paper on MEG signals detected with a SQUID magnetometer [7]. Some ten years later, the first multi-channel SQUID MEG systems emerged ([8] and references therein). Today, MEG is an increasingly important method in neuroscience and MEG systems with over 300 channels are commercially available (e.g. Elekta, Neuromag®).

MEG and EEG are particularly important because they are non-invasive, safe and they have a high temporal resolution of roughly 1 ms [9–11]. In contrast, functional

F. Öisjöen, *High-T$_c$ SQUIDs for Biomedical Applications: Immunoassays, Magnetoencephalography, and Ultra-Low Field Magnetic Resonance Imaging*, Springer Theses, DOI: 10.1007/978-3-642-31356-1_4, © Springer-Verlag Berlin Heidelberg 2013

magnetic resonance imaging (fMRI) systems have high spatial resolution (<1 mm) but poor temporal resolution (seconds) [12]. Positron emission tomography (PET) [13] is another important technique but the use of radioactive tracers violates the safety aspect.

In order to translate the recordings in MEG/EEG into localized sources of brain activity, one employs a source reconstruction technique meant to solve the "inverse problem". In order to simplify the inverse problem, the sources are typically modeled as current dipoles [10]. A current dipole can be thought of as a short segment of current with the return lines spreading far from the source current (see Sect. 4.2), similar to a magnetic dipole. The magnetic field lines from a current segment are circular around the direction of the current (right-hand rule). It is therefore easy to imagine that MEG is not sensitive to current dipoles aligned perpendicular to the sensor, typically referred to as radially aligned current dipoles. Radially in this case is referred to the direction perpendicular to the surface of the head if approximated as a sphere. EEG is more sensitive to the radially aligned current dipoles, thus, simultaneous MEG and EEG can provide discriminatory information for localizing the current dipoles [4, 14]. In order to localize the activity in the brain measured with MEG or EEG, the source localization made from the MEG/EEG data is mapped onto an MRI image. The electric field (that is measured by EEG) is distorted by the different electrical impedances of brain tissues, the skull and scalp. However, the magnetic fields are affected by the magnetic susceptibility that is roughly the same for different biological tissues. This is an advantage for MEG since it makes source localization more accurate [15, 16].

In EEG, electrodes are placed on the subject's scalp and the electrical potential between different electrodes is measured. From this regard, MEG is more practical for high number of channels since, contrary to EEG, the electrodes (SQUID channels) are already in place and do not need electrical contact. This advantage greatly reduces the subject preparation time.

In commercial MEG systems, low-T_c SQUIDs are successfully used due to their high reproducibility from a fabrication point of view, but also because of their high magnetic field resolution of 1–5 fT/$\sqrt{\text{Hz}}$ [9]. The SQUID devices are commonly made from thin films of niobium (Nb) with a superconducting transition temperature of 9 K. To reach this temperature, liquid helium is used as the cryogen. Thermal insulation requirements at these low temperatures place a lower limit for the separation between the cold sensor and the room temperature subject (e.g. a human head) to a few centimeters.

In contrast, by employing high-T_c devices cooled with liquid nitrogen, the separation between the sensor and the scalp can be reduced by more than one order of magnitude. Previous studies demonstrate the use of high-T_c SQUID devices for recordings of evoked potentials by auditory stimulus [17, 18] and electrical stimulus of the median nerve [19, 20] on human subjects. Tarte et al. [21] theoretically predicted the superior signal to noise available to a high T_c sensor compared to a low-T_c device. In a later publication, Magnelind et al. [22] demonstrated detection of the evoked magnetic field from *in vitro* rat brain hippocampal tissue slices using a high-T_c SQUID magnetometer.

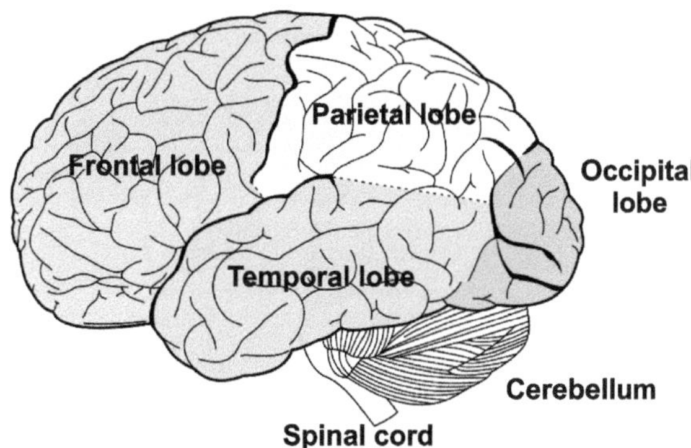

Fig. 4.1 The different parts of the cerebral cortex

Evoked potentials from humans are well-understood and enables averaging of thousands of events when recorded. This serves well as a proof-of-principle of MEG capability. Contrary to evoked potentials, spontaneous brain activities have to be measured without averaging which makes them challenging to detect. The focus of this thesis is measurements of spontaneous brain activity and not evoked activity that requires a stimulus. This is (to our knowledge) the first attempt to record spontaneous brain rhythms with high-T_c SQUIDs.

4.2 The Human Brain and Neurons

The human brain is divided into two hemispheres (left and right) by the longitudinal fissure. Brain activity associated with the most complex tasks is located in the outermost layer of cerebral cortex, also know as gray matter (2–4 mm thick sheet of tissue). This is also where most sources measured with MEG are located [10]. The cerebral cortex in each of the hemispheres, schematically shown in Fig. 4.1, is divided into four lobes: the frontal, parietal, temporal and occipital lobe. For additional information to brain physiology and anatomy, see [1, 23, 24]. These references [1, 23, 24] are also used in the following sections unless otherwise specified.

The Neuron

The building block for processing and transmission of information in the brain and the core of the nervous system is the neuron. There are roughly 100 billion (10^{11}) neurons in a human brain and they are interconnected through synapses. The number of synapses in a human brain is approximately 10^{18} which allows for a huge number of possible connections between neurons and is an indication of the great complexity of the human brain. A neuron consists of the soma which is where information is

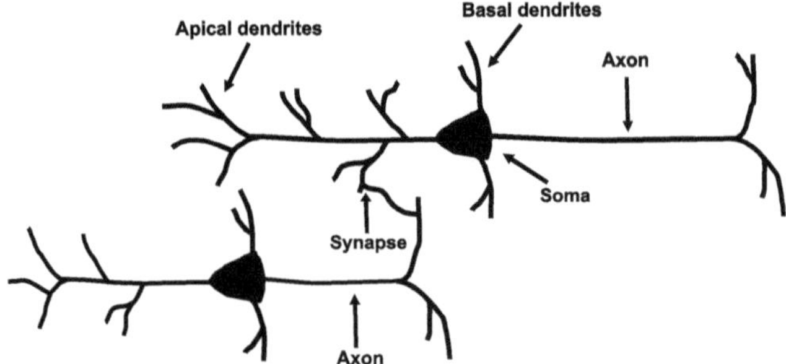

Fig. 4.2 A schematic drawing of two pyramidal neurons connected by a synapse. The apical dendrite is a long dendrite that branches many times as it extends from the soma. Basal dendrites are shorter and extrude directly from the soma. Also shown are the axons of the two neurons

processed, one axon and a number of dendrites that can branch hundreds of times (see Fig. 4.2). Several dendrites can extend from the soma but no more than one axon. The axon is responsible for transmission of signals to other neurons or distant parts of the body. Similar to dendrites, the axon can branch multiple times. A single axon is typically a few μm in diameter and can be as long as the height of the human body. In contrast, a dendrite receives signals from axons, and they are typically less than 1 mm long.

The resting potential across the cell membrane of a neuron is kept around -65 mV. Signal transmission through an axon is initiated by the generation of an action potential. The cell membrane is permeable to Na^+ and K^+ under certain conditions. As the action potential approaches, the local potential across the membrane changes and the Na^+ channels open. This allows for Na^+ to flow across the membrane and into the intracellular space. This process is called depolarization. The neuron is repolarized through deactivation of the Na^+ channels and activation of outflow of K^+ to the extracellular space. This process is repeated as the action potential travels towards the synapse. At the synapse, the Na^+ channels are opened on the dendrite and the post-synaptic current can be transmitted towards the soma.

Magnetic Field from a Current Dipole

A widely used concept for modeling in neuromagnetism is the current dipole representing a small patch of activated cortex with synchronously firing neurons. A current dipole may be thought of as a short segment of current where the current inside the neuron (the post-synaptic current in the dendrite) is known as the primary or impressed current. When a signal is being transmitted along the dendrite, the return path for the charges is through the extracellular space. These currents are called volume currents because they flow throughout the entire volume conductor (the head). This picture gives rise to the concept of a current dipole, illustrated in Fig. 4.3. The following derivation can be found in detail in [25].

Fig. 4.3 The concept of a current dipole showing the impressed current inside the neuron, the corresponding magnetic field lines, and the extracellular volume currents

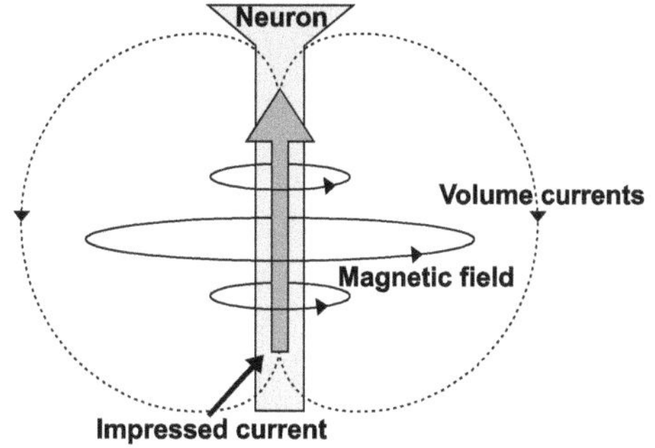

We can consider the impressed current density $\mathbf{J}^i$ to lie within a conductor G with conductivity σ and magnetic permeability $\mu \approx \mu_0$. The Maxwell's equations that describe the electric ($\mathbf{E}$) and magnetic field ($\mathbf{B}$) generated by the impressed current source $\mathbf{J}^i$ are given by

$$\mathbf{E} = -\nabla V \tag{4.1}$$

$$\nabla \times \mathbf{B} = \mu_0 \mathbf{J} \tag{4.2}$$

$$\nabla \cdot \mathbf{B} = 0 \tag{4.3}$$

$$\mathbf{J} = \mathbf{J}^i + \sigma \mathbf{E} \tag{4.4}$$

where V is the electric potential and $\mathbf{J}$ is the total current density. Note that the ohmic current $\sigma \mathbf{E}$ is the extracellular volume current. We insert $\mathbf{J}$ into the Biot-Savart law and find the magnetic field,

$$\mathbf{B}\left(\mathbf{r}\right) = \frac{\mu_0}{4\pi} \int_G \mathbf{J}(\mathbf{r}') \times \frac{\mathbf{r} - \mathbf{r}'}{|\mathbf{r} - \mathbf{r}'|^3} dv'. \tag{4.5}$$

Now, let us consider an unbounded homogeneous conductor, a simplified model of the human head with σ being constant in the entire conductor. These assumptions together with vector calculus and the Stoke's theorem yield

$$\mathbf{B}\left(\mathbf{r}\right) = \frac{\mu_0}{4\pi} \int_G \mathbf{J}^i\left(\mathbf{r}'\right) \times \frac{\mathbf{r} - \mathbf{r}'}{|\mathbf{r} - \mathbf{r}'|^3} dv'. \tag{4.6}$$

Here we see that in the case of a homogeneous unbounded conductor, the contribution to the magnetic field only comes from the impressed current and not from the volume currents. Now we make the impressed current a current dipole by concentrating it to a single point r_0, thus, $\mathbf{J}^i = \delta(r - r_0)\mathbf{Q}$ with moment $\mathbf{Q}$. Then we obtain

$$\mathbf{B}(\mathbf{r}) = \frac{\mu_0}{4\pi} \int_G \mathbf{J}^i(\mathbf{r}') \times \frac{\mathbf{r} - \mathbf{r}'}{|\mathbf{r} - \mathbf{r}'|^3} dv' \simeq \frac{\mu_0}{4\pi} \mathbf{Q} \times \frac{\mathbf{r} - \mathbf{r}_0}{|\mathbf{r} - \mathbf{r}_0|^3}, \qquad (4.7)$$

where $\mathbf{Q} = \int_G \mathbf{J}^i(\mathbf{r}')dv'$.

If we approximate the current dipole moment by $Q = \sigma A \Delta V$ [11], where $\sigma = 0.25\,\Omega^{-1}\,\mathrm{m}^{-1}$ is the conductivity of the dendrite, $A = 10^{-12}\,\mathrm{m}^2$ is the cross section area, and $\Delta V = 10\,\mathrm{mV}$ is the post-synaptic potential, we obtain $Q = 2 \times 10^{-15}\,\mathrm{Am}$. We insert this number into Eq. 4.7 and calculate the magentic field 5 cm from the source to be roughly 10^{-19} T. This is six orders of magnitude smaller than is typically measured with MEG. This simple calculation gives an indication of how many neurons have to fire synchronously, and be aligned (corresponding to 1,000,000 synapses) to produce a measurable signal outside the head.

Spontaneous Brain Activity

Brain activity that occurs without any externally applied stimuli such as sensory input or motor output are classified as spontaneous brain activities. Spontaneous brain activity is important for studies of brain development and functions [26], as well as sleep research [27, 28], but also for research related to e.g. Alzheimer's disease [29] and Parkinson's disease [30]. The spontaneous brain activities are classified into different frequency ranges: 0–4 Hz (δ), 4–8 Hz (θ), 8–13 Hz (α), 13–30 Hz (β), and 30–100 Hz (γ).

The δ-activity is strongest during sleep in the frontal lobes of adults or in the posterior regions in infants [27, 28].

The θ-rhythm (4–8 Hz) is typically stronger in children and tends to decay in amplitude with age [31]. It is usually strongest over the frontal lobes and is typically associated with working memory [31–34] or mental tasks such as calculations [14]. It is also seen during meditation and drowsiness in adults.

The α-activity is the most well known of the spontaneous brain rhythms. It is present in the 8–13 Hz range and is located predominantly in the occipital region of the cortex but also in the parietal region close to the occipital lobe [26, 35] (see Fig. 4.1). α-activity is strong during relaxed wakefulness with closed eyes, and is attenuated when one opens his/her eyes [26, 35]. It is believed to come from neurons in the visual cortex being synchronized in the absence of visual input, however, visual imagery [36] and mental tasks [31, 37] have been shown to attenuate the α amplitude, even with eyes closed. The frequency of the α-rhythm is well known to change with age [31]. Some subjects do not show any suppression of the α-rhythm when the eyes are open and some show only weak α-rhythms [35].

Another oscillatory rhythm, present in the same frequency range as the α-activity is the μ-activity. This brain rhythm is a resting potential of the motor cortex (and therefore found in that area, roughly between the parietal and frontal lobes) and is attenuated with motor activation such as moving a limb or flexing the hands [26].

The β-rhythm is found symmetrically in both hemispheres. It is typically detected in an alert state during active concentration and is related to control of motor functions [38].

The final rhythm is in the higher frequency range. The γ-rhythm is presently as subject of debate and is somewhat controversial. For a review on the topic see [39].

4.3 Disturbances and Artifacts

Disturbances to the MEG signal can be both external, coming from the environment or the experimental setup itself, or internal/biological, coming from the subject. A magnetically shielded room suppresses external magnetic sources of noise efficiently. It typically consists of several layers of high-permeability mu-metal (e.g. an iron-nickel alloy) and at least one layer of aluminum to shield from high frequency noise. The high-permeability mu-metal provides a low reluctance path for the magnetic field lines to go around the room.

Examples of non-biological sources that can cause distortion of MEG signals are mechanical vibrations, power line interference (50/60 Hz), elevators or other large metallic objects in motion. The frequency component of large moving objects are typically well below 1 Hz [1].

Artifacts coming from the subject are usually more influential and can come from heart beats, eye movements, eye blinks, movements, breathing and muscle activity [1]. Breathing and movements typically appear as low frequency drift below 1 Hz. The strength of the magnetic field component of a current dipole depends strongly on orientation and the distance and calculate the magentic field. Therefore, head movements are influential in MEG since the sensor positions are not fixed on the head. This type of artifact can be completely eliminated with a high-T_c MEG system, as discussed in Sect. 6.1. In a conventional MEG system, localization coils on the head are used to find the head position [1]. Signals originating from muscle activity (e.g. neck or facial muscles) are typically reflected over a wide range of frequencies (typically above 10 Hz) with high amplitudes [40, 41]. The simplest way to reduce the effect of muscle artifacts is to use a low-pass filter or to reject segments containing the high amplitude artifacts. Artifacts from eye movements or blinks are more prominent in the frontal channels.

The strength of the magnetic field produced by the QRS complex of the heartbeat (the part of the cycle when the heart contracts) can be several tens of pT if measured on the chest with a SQUID magnetometer, as shown in Fig. 4.14 (left panel). The magnetic field produced by the cardiac muscle can strongly influence the MEG recording, especially on SQUID magnetometer channels. One can reduce this source of artifact by e.g. using SQUID gradiometers or by averaging the signal [10]. Often, one records the heartbeat with ECG (electrocardiography) simultaneously as the MEG. This way the segments containing the strong heartbeat artifacts can be rejected, or stimulus can be applied between heartbeats.

4.4 Experimental Methods

In this section, experimental details of the MEG recordings are described. MEG recordings were initially made with a single channel (one SQUID sensor). This serves well as a proof-of-principle. The latest upgrade was the extension to a two-channel system. The experiments have been focused on detecting the α-rhythm from the occipital part of the cortex and the μ-rhythm from the motor cortex of human subjects. We have also found an anomalous θ-band activity discussed later. At times, performed simultaneous EEG with our MEG. The details regarding the experiments are presented in the following sections.

4.4.1 Experimental Setup and Protocol

The magnetic signals produced by neural currents in the human brain are extremely weak, therefore, MEG recordings require magnetic shielding. All the MEG recordings presented were performed inside a magnetically shielded room with 2-layers of mu-metal sandwiching a copper-coated aluminum layer (Vacuum Schmeltze GmbH, located at Imego AB). The copper-coated aluminum shields the higher frequency components. The shielding factor at 10 Hz is roughly 10,000, according to specifications from the manufacturer and the background dc magnetic fields was measured to be roughly 40 nT (only measured along one axis).

The cryostats used for the MEG recordings were of the same type as the one described in Sect. 3.3.1. Briefly, it was made of non-magnetic glass fiber reinforced epoxy. The SQUID sensor was glued onto a sapphire rod that was thermally connected to a liquid nitrogen bath (0.7 l of liquid nitrogen). One critical feature of the cryostat was the possibility to manually move the cold stage arbitrarily close to the 200 μm thick sapphire window that separated the cold SQUID sensor and a subject's scalp (room temperature). This enabled a SQUID-to-scalp separation of less than 1 mm, see Fig. 4.4.

A PC controlled the SQUID electronics (Magnicon SEL-1) and LabVIEW programs were developed for controlling other hardware units used. The MEG recordings were acquired either with a spectrum analyzer (Stanford SR785) or with a data acquisition card (NI-DAQ 6014 or NI-DAQ 6251). The spectrum analyzer method was not satisfying since it can only provide frequency information. With the DAQ, the output of the SQUID electronics was sent to a pre-amplifier (Stanford SR560, typically with a gain of 1–10 K and a bandpass filter at 1–30 Hz) and then to the DAQ. The data was most commonly further filtered in the LabVIEW program.

For recordings in the single-channel setup, the cryostat was placed behind a seated subject, schematically shown in Fig. 4.4. The subject relaxed his/her head against a pillow (not shown) with a hole in it where the cryostat could approach the subject's scalp. The surface of the sapphire window of the cryostat was kept flat against the subject's head in order to further reduce the SQUID-to-scalp distance. Recordings

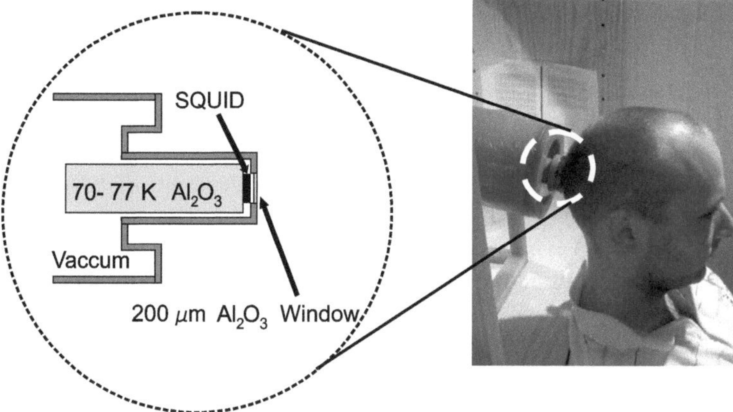

Fig. 4.4 A simple schematic showing a drawing of the uppermost part of the cryostat and the position of the cryostat relative to the subject's head during recordings in the single-channel setup. The cold SQUID and the room-temperature head are separated by a $200\,\mu$m thick sapphire window

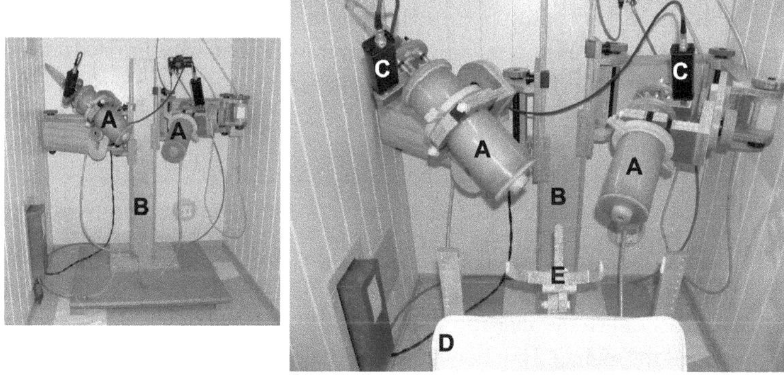

Fig. 4.5 Photographs showing the two-channel high-T_c MEG setup inside the magnetically shielded room. *Left* The entire system with (*A*) two liquid nitrogen cryostats mounted on (*B*) the two flexible arms of the wooden stand. *Right* A close up of the system with (*A*) the cryostats, (*B*) stand, (*C*) SQUID electronics, (*D*), subject chair, and (*E*) the u-shaped headrest

were made with either eyes open or eyes shut through the entire recording, or, the subject was verbally instructed to open or close his/her eyes during the recording.

Since the target of the single-channel recordings was the α-rhythm, we recorded from the occipital region of the brain in the O1 or O2 position in the 10–20 international reference system. Generally speaking, the α-rhythm is strongest close to O1–O2 (occipital region). However, the α-rhythm is also detected in other locations close to this region. In order to verify this with our system, and also to maximize the signal strength, we first tried recording from different positions on the scalp with eyes closed.

4.4.2 Two-Channel MEG

The single channel system that was initially used was extended with an additional channel. This was done by adding another cryostat, with the same design as the one previously described, and with a SQUID magnetometer operated with its own electronics (Magnicon SEL-3). In order to increase the flexibility of the setup, a stand was constructed for the two cryostats as shown in Fig. 4.5. The stand, made in wood to ensure there was no magnetic material close to the SQUID, had two arms holding the cryostats that could be moved to arbitrary locations on the scalp of a seated subject. The only limitation was the size of the cryostats that limited the distance between measurement locations to roughly 10 cm. With this setup it was possible to correlate brain activity from different parts of the brain or one SQUID could be used as a reference for measuring the background noise simultaneously as the brain signal was recorded. The data acquisition was made with the same NI-DAQ 6251 preceded by two (one for each channel) Stanford pre-amplifiers (SR560) as described in the previous section. To stabilize the head position a u-shaped headrest was constructed, as shown in Fig. 4.5. The headrest was flexible in order to fit different head sizes and to be able to fix the position of the head.

With the two-channel setup we measured the μ-rhythm found in the motor cortex (C3–C4) as well as the occipital α-rhythm. The μ-rhythm was attenuated with motor activation (e.g. flexing the hands). Instructions (eyes open/closed, hands flexed/relaxed) were verbally given to the subjects.

4.5 Results and Discussion

Conventional MEG systems incorporate low-T_c SQUIDs in a helium dewar with a typical spacing of more than 2 cm between the cold SQUID and the room temperature subject's scalp. In order to estimate the higher signal available to a high-T_c SQUID due to the reduced SQUID-scalp spacing compared to a low-T_c system, a current dipole was simulated. With the simulations, we get a ratio of the signal available to a high-T_c and a low-T_c SQUID sensor from sources at different depth's in the brain.

With the 2-channel setup we recorded the μ-rhythm simultaneously as the α-rhythm. The μ-rhythm was modulated by flexing the hands.

4.5.1 Simulations

The model used for the simulations was a single point source current dipole in a homogeneous volume conductor. This is a simple model but it gives an indication of how much stronger the signal available to a high-T_c device is compared to a low-T_c device in typical systems of the corresponding technology. In the simulations, one

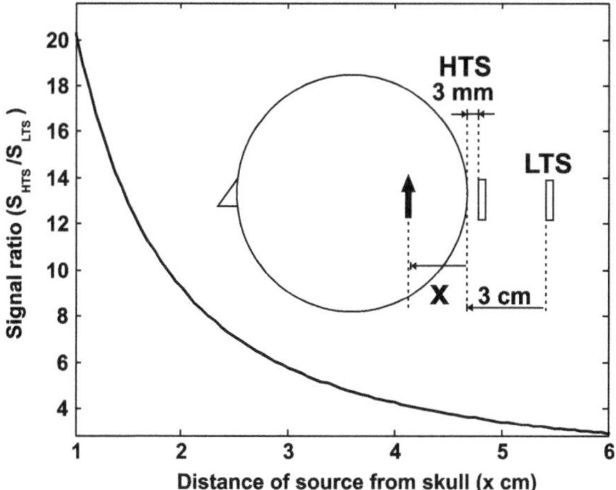

Fig. 4.6 The ratio between the magnetic field strength available to a high-T_c SQUID (HTS) and low-T_c SQUID (LTS) with their typical stand-off. The current dipole point source indicated by the black arrow in the inset, is moved from 1 to 6 cm inside the skull. The magnetic field is integrated over the size of the sensors placed 3 mm (HTS) and 3 cm (LTS) from the skull. For shallow sources (1 cm from the skull), the strength of the magnetic field is roughly 20 times stronger for the high-T_c SQUID as compared with the low-T_c SQUID

Fig. 4.7 A map of the scalp of a human head seen from above. Four MEG recordings are included measured at the indicated locations on this subject. The highest signal in the α-range of activity is present at 10.5 Hz, measured to *the left* in the occipital region (roughly O1)

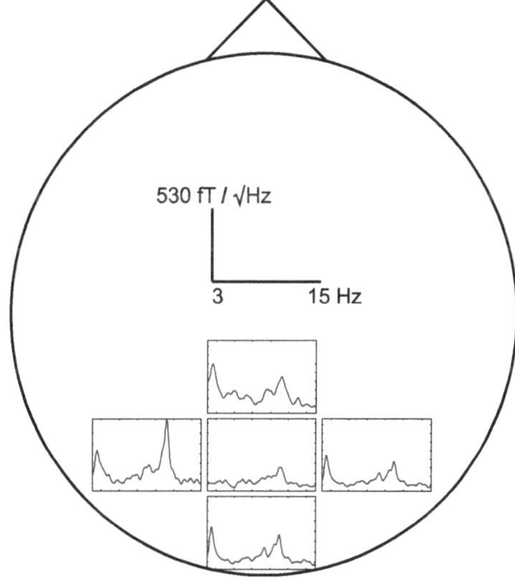

$10 \times 10\,\text{mm}^2$ sensor was placed 3 mm from the scalp (high-T_c) and another 3 cm from the scalp (low-T_c). The current dipole was placed at different distances from the scalp starting at 1 cm (shallow sources in brain) going to deeper sources at a maximum depth of 6 cm. The signal was integrated over the area of the sensors and the ratio of the signals were calculated and plotted (Fig. 4.6). As seen in Fig. 4.6, for

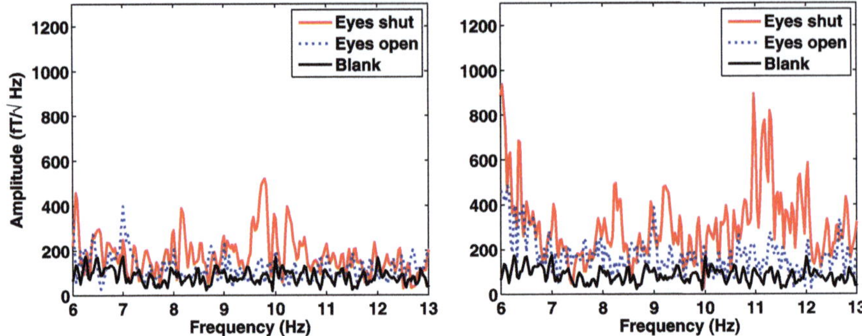

Fig. 4.8 MEG recordings on different subjects measured with a spectrum analyzer. In both cases, the activity in the α-range is attenuated when the subject has open eyes (subject 1, left, at 10 Hz and subject 2, right, at 11 Hz). This is indicative of the α-rhythm

sources on the surface of the cerebral cortex, closest to the skull, the magnetic field strength available to the high-T_c sensor is 20 times higher than for a typical low-T_c SQUID sensor. For deeper sources at 5 cm, the ratio is more than 3.

4.5.2 Brain Activity Recordings

The first MEG recordings made with our high-T_c SQUIDs were performed with a simple single-channel setup. These results will be shown first and are focused on the α-rhythm. Subsequently, α- and μ-rhythm recordings with a more robust two-channel setup will be presented. Lastly, a discussion about a suggested high amplitude θ-range rhythm in the occipital region, detected with our setup will follow.

α and μ-Rhythms

The map of recordings in Fig. 4.7 shows activity in the 3–15 Hz range measured with our high-T_c system using the MAG3R sensor. The recordings were measured sequentially and not simultaneously since this setup only employed a single MEG channel. However, Fig. 4.7 clearly indicates where the highest amplitude of α-activity can be measured. The highest signal was obtained in the O1 location and the amplitude was roughly 500 fT/$\sqrt{}$Hz at 10.5 Hz. The 30 s recordings were sampled at 1,070 Hz and amplified and filtered (gain $=$ 100, bandpass filter 1–100 Hz).

A well known feature of the α-rhythm is the attenuation of the activity with eyes open, known as the Berger effect. In order to detect this, we measured 2 traces: one with eyes open and one with eyes closed. In addition, we recorded a blank trace to see the full system noise. We recorded the two different traces (eyes open and eyes closed) on two subjects with a spectrum analyzer (1–25 Hz). The recorded data

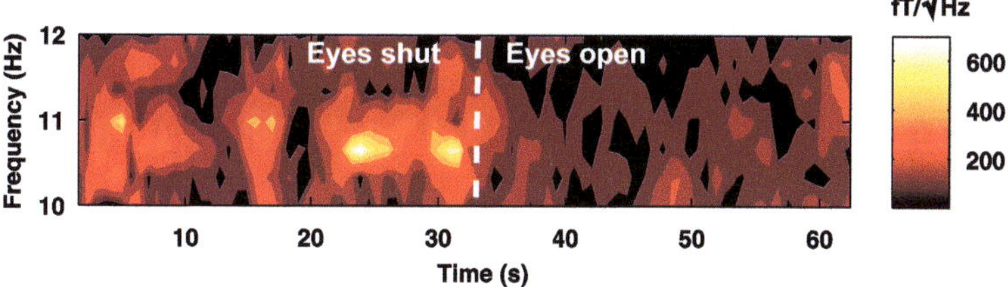

Fig. 4.9 Spectrogram of a single channel MEG recording from the occipital region of a subjects head. The subject was instructed to have eyes closed during the first 32 s. After being verbally instructed, the subject opened his eyes after 32 s. The α-activity present at 10.5–11 Hz is, as expected, attenuated with eyes open

is plotted in Fig. 4.8. In subject 1 (left), activity centered around 10 Hz is clearly attenuated when the eyes are open. The corresponding phenomenon was observed in subject 2 (right) but at a slightly higher frequency ($\sim$11 Hz). The amplitude of the α-rhythm of subject 2 is about 800 fT/$\sqrt{\text{Hz}}$.

To investigate the nature of the activities we observed, we recorded time resolved MEGs. A 64 s recording is shown in Fig. 4.9. The subject started with eyes closed and was verbally instructed to open them at 32 s as indicated by the dashed line in the figure. This figure was generated by performing an FFT on e.g. the first 3 s window of the time-domain data and plot it at time 1.5 s (the first point). Then the window moved 0.7 s and 3 s window is FFT'd (thus, the windows are overlapping) and plotted it at 2.2 s, etc. This generated a spectrogram with the frequency on the y-axis, the time on the x-axis, and the amplitude is plotted as a color coded contour map. The plot shows (Fig. 4.9) the upper range of the α-band and the attenuation of the activity between 10.5–11 Hz with eyes open is obvious. The peak amplitude is around 600 fT/$\sqrt{\text{Hz}}$ with eyes closed and is attenuated to the system noise with eyes open ($<$50 fT/$\sqrt{\text{Hz}}$).

The single-channel setup was upgraded to a two-channel MEG as previously described. The addition of the second cryostat and the support structures did not add any additional noise contribution.

With this new configuration, more reliable and reproducible placement of the SQUID sensors was possible, and therefore the quality of the recordings improved. In Fig. 4.10 a recording from the occipital region (O2) of a human subject is shown. In this trace, the suppression of the amplitude of the α-recording, bandpass filtered (8–13 Hz), is evident in the time trace. Note that this is without any averaging of the signal, unlike those recorded evoked signals that were averaged many times (100–10,000) to improve SNR previously made with high-T_c SQUIDs [17–20]. Moreover, in this figure (lower panel), one can also see a lower band α-rhythm at 9 Hz, which was confirmed by EEG.

With the two-channel MEG setup we were able to measure brain activity from two locations on the scalp simultaneously. This led us to attempt simultaneous recordings of the occipital α-rhythm and the μ-rhythm found in the motor cortex, both in the 8–13 Hz range. For this experiment, one SQUID channel was placed in the occipital

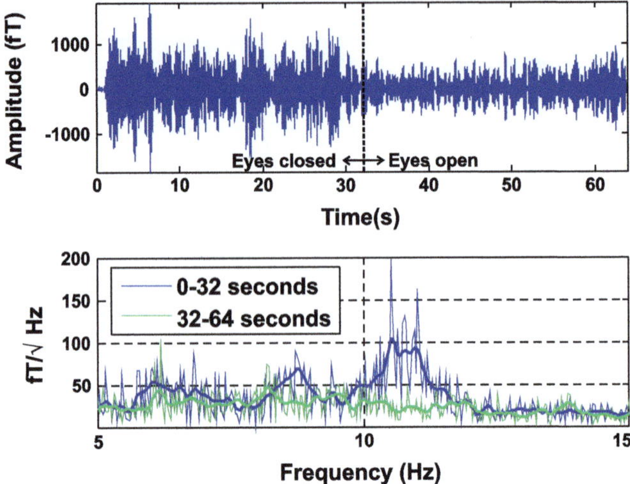

Fig. 4.10 A 64 s recording from the occipital region of an awake and relaxed human subject. The subject had his eyes closed for the first 32 s and open for the last 32 s. *The upper panel* (adapted from [42]) shows the raw data bandpass filtered from 8–13 Hz. A clear suppression of the signal can be seen as the subject opens the eyes after 32 s. *The lower panel* shows the Fourier transform of the two time intervals (0–32 s and 32–64 s). The detected peak at 10.6 Hz is attenuated with open eyes. A weaker α-rhythm can also be seen at 9 Hz. The Fourier transformed traces have been smoothed (*thick lines*) using Gaussian window averaging

region (O2) and the other SQUID above the motor cortex (C4, roughly above the right ear). The subject was instructed to start with eyes open and hands relaxed and after 32 s, was instructed to close the eyes and flex the hands. The expected outcome was that for the first 32 s, the μ-rhythm should be present in the SQUID placed in C4, and the α-rhythm should be attenuated in O2 and vice versa in the last 32 s. The measured data is presented as two spectrograms in Fig. 4.11. Indeed, we see a μ-rhythm in the C4 channel (upper panel) at about 11 Hz present in the first half of the recording. Furthermore, the α-rhythm in O2 (lower panel) is attenuated in this time frame. After the instructions given at 32 s (close the eyes and flex the hands), the α-rhythm is strong and the μ-rhythm is attenuated in Fig. 4.11, as expected.

Simultaneous EEG and MEG Recordings

In order to verify the recordings with our system, we recorded EEG simultaneously with the high-T_c MEG. EEG is a well established method for recording electrical signals from the brain and served primarily as a verification method for the α- and μ rhythm recording but also for investigations regarding other activities such as the θ-rhythm. We used a 32-channel system with sampling frequency of 1 kHz. The SQUIDs were placed as close as possible to the EEG electrodes at e.g. O2 or C4. The presence of the electrodes did not add any noticeable noise to the SQUID recordings.

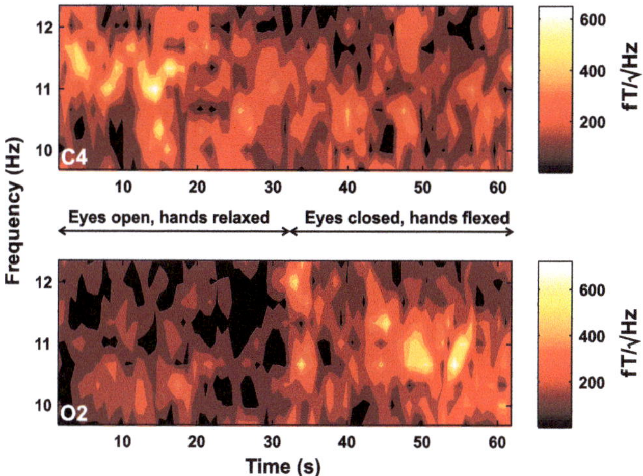

Fig. 4.11 Two recordings made simultaneously with the two MEG channels at two different locations on the head. *The upper panel* shows the recording from the motor cortex (C4) and *the lower panel* shows the occipital channel (O2). Adapted from [42]

The spectrogram in Fig. 4.12 shows two recordings made simultaneously with our MEG (upper panel) and the EEG (lower panel) from the occipital region (O2) of a human subject's head. The α-rhythm is seen in both recordings at 11–11.5 Hz during roughly the first 25 s when the subject had eyes closed. In the rest of the recording the subject had open eyes and the α-rhythm was attenuated in both MEG and EEG.

Anomalous θ-Rhythm

During many of our experiments we detected a signal in the θ-range (4–8 Hz) from the occipital region of the head. What initially made us interested in this signal is the high amplitude in the θ-range in this part of the brain, not previously observed in state-of-the-art MEG nor EEG. Figure 4.13 shows the spectrogram of an MEG recording from the occipital region of one subject. It shows the α recording of Fig. 4.9 in the upper panel, but it also includes the θ-band in the lower panel. The θ-activity seen at 6.5 Hz shifted to about 6.7 Hz when the subject had open eyes. The high amplitude of this rhythm in this part of the brain is atypical for state-of-the-art MEG.

Sources of brain activity inside the brain are typically considered to be current dipoles. This model was used in the simulations and is necessarily simple for practical purposes. In a real situation, the head is not spherical, homogeneous and unbound which makes modeling more complicated. In particular, a source inside the brain is not necessarily a simple point source current dipole. A small piece of the cortex contains a huge number of active neurons that may or may not be active simultaneously. Their magnetic signatures are thus likely to include higher order sources of magnetic activity. However, the magnetic field from such higher order sources decays even

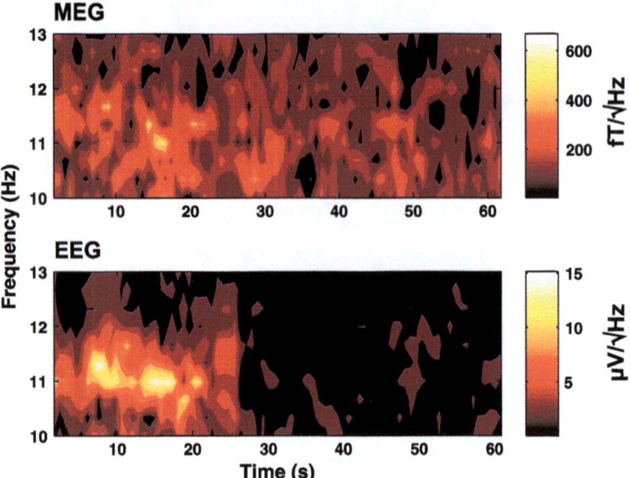

Fig. 4.12 Spectrograms of simultaneously recorded EEG and high-T_c SQUID MEG from the occipital region (O2) of a human subject's head. The α-rhythm, present in both recordings at 11–11.5 Hz during the first 25 s was attenuated when the subject had open eyes (time 25–64 s) in both channels

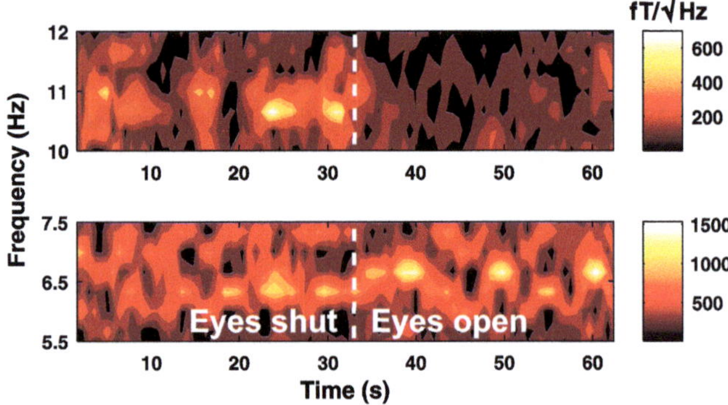

Fig. 4.13 Spectrogram of a single-channel MEG recording from the occipital region of the head. *The upper panel* shows the higher range of the α-band and the 10–11 Hz signal is attenuated with open eyes (same as Fig. 4.9). *The lower panel* shows the recorded signal in the upper θ-band. The amplitude of the activity at 6.5–6.7 Hz is roughly twice that of the α-signal in *the upper panel*

more rapidly than that of a dipole (i.e. quadrupoles decay as $1/r^5$ and octupoles as $(1/r^7)$ [43]). The amplitude of the higher order sources a few centimeters away from the scalp may be too low for the low-T_c SQUIDs in conventional MEG systems to detect. Therefore, the proximity of the high-T_c SQUIDs to the sources may be responsible for the high θ-range amplitudes detected in the occipital region. Furthermore, this anomalous θ-band activity was not detected by the EEG. This may be due to the alignment of the sources in the brain since EEG is less sensitive to tangentially

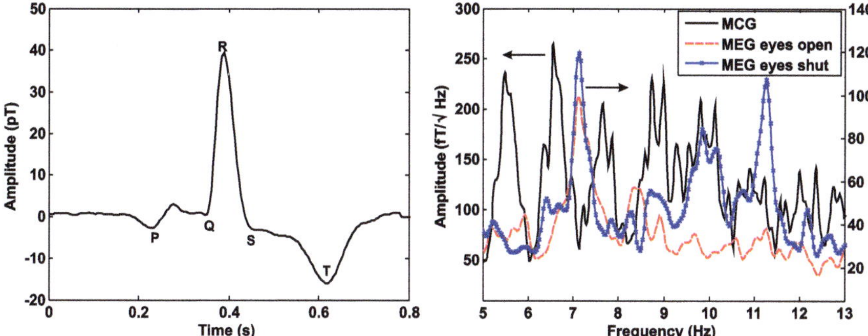

Fig. 4.14 *The left panel* shows one full cycle of the magnetic signal produced by a human heartbeat measured with a high-T_c SQUID magnetometer. This trace is an average of roughly 120 heartbeats and shows the different parts of the PQRST complex. The amplitude at the R-peak is 40 pT. *The right panel* shows the frequency spectrum of the magnetic signal measured by one MCG channel and one MEG channel placed in O2. Note that the 7 Hz signal in the MEG channel is located between two peaks of the MCG signal

(with respect to the scalp) aligned current dipoles. Another possible explanation is if the nature of the source is of multi-pole type which may be difficult for EEG to detect.

Although we were not able to reliably modulate the discovered occipital θ-range activity, we noticed that the frequency sometimes shifted as eyes were opened or closed. This may suggest a mechanism for modulation of this activity is related to e.g. attention or awareness. The unknown behaviour of this brain activity made it difficult to investigate the possibility of a higher order source. Due to the rather unreliable amplitude of brain sources in general, it is difficult to make a simple source-to-sensor experiment and plot the amplitude versus distance. One proposed way to investigate the nature of the anomalous θ-rhythm is to compare the relative amplitude of this signal to the α-rhythm at different distances. This should indicate whether there is a significant difference in amplitude versus distance between the α and the θ-rhythms. For example, if the θ-rhythm is higher in amplitude than the α-rhythm close to the scalp and the relative difference (or ratio) between the different signals is not the same further away may indicate different nature of the two sources.

Investigation of Correlation with Cardiac Magnetic Signals

In order to investigate how the heartbeats of subjects affect our MEG recordings, we placed one SQUID close to the chest and the other in O2 and performed simultaneous MEG and magnetocardiography (MCG). With MCG, we can find the magnetic signature of the heartbeat, shown in the left panel of Fig. 4.14. This is particularly important for our θ-rhythm findings, to ensure that they are not cardiac artifacts. In Fig. 4.14 (right panel) the frequency spectrum of an MEG recording from O2 with eyes open (red, dashed line) and eyes closed (blue, crosses), and the MCG (black) recorded simultaneously are shown. We clearly see harmonic components of the heartbeat at e.g. 5.5, 6.5, and 7.6 Hz in the MCG channel. More importantly, the θ-rhythm at 7 Hz is located in between two of the harmonics of the heartbeat. We can also see that the α-rhythm at 11 Hz is suppressed with eyes open, as expected. From this measurement we concluded that the θ-rhythm we recorded was not a cardiac artifact.

References

1. P.C. Hansen, M.L. Kringelbach, R. Salmelin (eds.), *MEG: An Introduction to Methods* (Oxford University Press, New York, 2010)
2. R. Caton, The electric currents of the brain. Br. Med. J. **2**, 278 (1875)
3. H. Berger, Über das electrenphalogramm des Menschen. Arch. Physchiatr. Nervenkr. **87**, 527–570 (1929)
4. G.L. Barkley, C. Baumgartner, MEG and EEG in epilepsy. J. Clin. Neurophysiol. **20**, 163–178 (2003)
5. D. Cohen, Magnetoencephalography: Evidence of magnetic fields produced by alpha-rhythm currents. Science **161**, 784–786 (1968)
6. R.C. Jaklevic, J. Lambe, A.H. Silver, J.E. Mercereau, Quantum interference effects in Josephson tunneling. Phys. Rev. Lett. **12**, 159–160 (1964)
7. D. Cohen, Magnetoencephalography: detection of the brain's electrical activity with a super-conducting magnetometer. Science **175**, 664–666 (1972)
8. G.L. Romani, P. Rossini, Neuromagnetic functional localization: principles, state of the art, and perspectives. Brain Topogr. **1**, 5–21 (1988)
9. A.I. Ahonen, M.S. Hämäläinen, M.J. Kajola, J.E.T. Knuutila, P.P. Laine, O.V. Lounasmaa, L.T. Parkkonen, J.T. Simola, C.D. Tesche, 122-channel SQUID instrument for investigating the magnetic signals from the human brain. Phys. Scr. **T49**, 198–205 (1993)
10. M. Hämäläinen, R. Hari, R.J. Ilmoniemi, J. Knuutila, O.V. Lounasmaa, Magnetoencephalography: theory, instrumentation, and applications to noninvasive studies of the working human brain. Rev. Mod. Phys. **65**, 413–497 (1993)
11. C. del Gratta, V. Pizzella, F. Tecchio, G.L. Romani, Magnetoencephalography—a noninvasive brain imaging method with 1 ms time resolution. Rep. Prog. Phys. **64**, 1759–1814 (2001)
12. N.K. Logothetis, J. Pauls, M. Augath, T. Trinath, A. Oeltermann, Neurophysiological investigation of the basis of the fMRI signal. Nature **412**, 150–157 (2001)
13. M.E. Phelps, J. Hoffman, N.A. Mullani, M.M. Ter-Pogossian, Application of annihilation coincidence detection to transaxial reconstruction tomography. J. Nucl. Med. **16**, 210–224 (1975)
14. K. Iramina, S. Ueno, S. Matsuoka, MEG and EEG topography of frontal midline theta rhythm and source localization. Brain Topogr. **8**, 329–331 (1996)
15. Y.C. Okada, A. Lahteenmäki, C. Xu, Experimental analysis of distortion of magnetoencephalography signals by the skull. Clin. Neurophysiol. **110**, 230–238 (1999)
16. R.M. Leahy, J.C. Moasher, M.E. Spencer, M.X. Huang, J.D. Lewine, A study of dipole localization accuracy for MEG and EEG using a human skull phantom. Electroenceph. Clin. Neurophysiol. **107**, 159–173 (1998)
17. M.S. Dilorio, K.Y. Yang, S. Yoshizumi, Biomagnetic measurements using low-noise integrated SQUID magnetometers operating in liquid nitrogen. Appl. Phys. Lett. **67**, 1926–1928 (1995)
18. Y. Zhang, Y. Tavrin, M. Mück, A.I. Braginski, C. Heiden, S. Hampson, C. Pantev, T. Elbert, Magnetoencephalography using high temperature rf SQUIDs. Brain Topogr. **5**, 379–382 (1993)

19. H.J. Barthelmess, M. Halverscheid, B. Schiefenhovel, E. Heim, M. Schilling, R. Zimmermann, Low-noise biomagnetic measurements with a multichannel dc-SQUID system at 77 K. IEEE. Trans. Appl. Supercond. **11**, 657–660 (2001)
20. D. Drung, F. Ludwig, W. Muller, U. Steinhoff, L. Trahms, H. Koch, Y.Q. Shen, M.B. Jensen, P. Vase, T. Holst, T. Freltoft, G. Curio, Integrated $YBa_2Cu_3O_{7-x}$ magnetometer for biomagnetic measurements. Appl. Phys. Lett. **68**, 1421–1423 (1996)
21. E.J. Tarte, P.E. Magnelind, A.Y. Tzalenchuk, A. Lõhmus, D.A. Ansell, M.G. Blamire, Z.G. Ivanov, R.E. Dyball, High T_c SQUID systems for magnetophysiology. Phys. C **368**, 50–54 (2002)
22. P.E. Magnelind, D. Winkler, E. Hanse, E.J. Tarte, Magnetophysiology of Brain slices using an HTS SQUID magnetometer system, in V. In, P. Longhini, A. Palacios (eds.) *Applications of Nonlinear Dynamics, Understanding Complex Systems* (Springer, Berlin, 2009), pp. 323–330
23. J.F. Stein, C.J. Stoodley, *Neuroscience an Introduction* (Wiley, Chichester, 2006)
24. J. Malmivuo, R. Plonsey, *Bioelectromagnetism: Principles and Applications of Bioelectric and Biomagnetic Fields* (Oxford University Press, New York, 1995)
25. J. Sarvas, Basic mathematical and electromagentic concepts of the biomagnetic inverse problem. Phys. Med. Biol. **32**, 11–22 (1987)
26. R. Hari, R. Salmelin, Human cortical oscillations: a neuromagnetic view through the skull. Trends Neurosci. **20**, 44–49 (1997)
27. W.J. Lutter, M. Maier, R.T. Wakai, Development of MEG sleep patterns and magnetic auditory evoked responses during early infancy. Clin. Neurophysiol. **117**, 522–530 (2006)
28. N.R. Simon, I. Mansheden, F.H. Lopes da Silva, A MEG study of sleep. Brain Res. **860**, 64–76 (2000)
29. E. Zamrini, F. Maestu, E. Pekkonen, M. Funke, J. Makela, M. Riley, R. Bajo, G. Sudre, A. Fernandez, N. Castellanos, F. del Pozo, C.J. Stam, B.W. van Dijk, A. Bagic, J.T. Becker, Magnetoencephalography as a putative marker for Alzheimer's disease. Int. J. Alzheimers Dis. **280289**, 2011 (2011)
30. J.L.W. Bosboom, D. Stoffers, C.J. Stam, B.W. van Dijk, J. Verbunt, H.W. Berendse, E. Ch. Wolters, Resting state oscillatory brain dynamics in Parkinson's disease: an MEG study. Clin. Neurophysiol. **117**, 2521–2531 (2006)
31. W. Klimesch, EEG alpha and theta oscillations reflect cognitive and memory performance: a review and analysis. Brain Res. Rev. **29**, 169–195 (1999)
32. S. Raghavachari, M.J. Kahana, D.S. Rizzuto, J.B. Caplan, M.P. Kirschen, B. Bourgeois, J.R. Madsen, J.E. Lisman, Gating of human theta oscillations by a working memory task. J. Neurosci. **21**, 3175–3183 (2001)
33. C.D. Tesche, J. Karhu, Theta oscillations index human hippocampal activation during a working memory task. Proc. Natl. Acad. Sci. USA **97**, 919–924 (1999)
34. M.J. Kahana, D. Seelig, J.R. Madsen, Theta returns. Curr. Opin. Neurobiol. **11**, 739–744 (2001)
35. C. Ciulla, T. Takeda, H. Endo, MEG characterization of spontaneous alpha rhythm in the human brain. Brain Topogr. **11**, 211–222 (1999)
36. S. Salenius, M. Kajola, W.L. Thompson, S. Kosslyn, R. Hari, Reactivity of magnetic parieto-occipital alpha rhythm during visual imagery. Electroenceph. Clin. Neurophysiol. **95**, 453–462 (1995)
37. H. Petsche, S. Kaplan, A. von Stein, O. Filz, The possible meaning of the upper and lower alpha frequency ranges for cognitive and creative tasks. Int. J. Phsychophysiol. **26**, 77–97 (1997)
38. S.N. Baker, Oscillatory interactions between sensorimotor cortex and the periphery. Curr. Opin. Neurobiol. **17**, 649–655 (2007)
39. J.R. Hughes, Gamma, fast, and ultrafast waves of the brain: their relationships with epilepsy and behaviour. Epilepsy Beh. **13**, 25–31 (2008)
40. J.M. Kilner, S.N. Baker, S. Salenius, R. Hari, R.N. Lemon, Human cortical muscle coherence is directly related to specific motor parameters. J. Neurosci. **20**, 8838–8845 (2000)

41. J. Mellinger, G. Schalk, C. Braun, H. Preissl, W. Rosenstiel, N. Birbaumer, A. Kübler, An MEG-based brain-computer interface (BCI). Neuroimage **36**, 581–593 (2007)
42. F. Oisjöen, J. F. Schneiderman, G. A. Figueras, M. L. Chukharkin, A. Kalabukhov, A. Hedström, M. Elam, and D. Winkler. High-T_c superconducting quantum interference device recordings of spontaneous brain activity: Towards high-T_c magnetoencephalography. Appl. Phys. Lett., 100(132601), 2012. 10.1063/1.3698152.
43. K. Jerbi, J.C. Mosher, S. Baillet, R.M. Leahy, On MEG modelling using multipolar expansions. Phys. Med. Biol. **47**, 523–555 (2002)

Chapter 5
Ultra Low Field Magnetic Resonance Imaging

In this chapter the ongoing effort to develop a system for magnetic resonance imaging in ultra-low magnetic fields is discussed. The following sections will introduce the concepts of the technique along with some background. The present status of our system will be described along with some preliminary results.

5.1 Introduction

Magnetic resonance imaging (MRI) is a widely used technique for non-invasive imaging of internal organs such as the human brain, as shown in Fig. 5.1. It was first described in 1973 [1] and today there are roughly 25,000 MRI machines worldwide. MRI is based on detection of precessing magnetic moments of protons in an external magnetic field, this is also known as nuclear magnetic resonance (NMR) [2, 3]. An MRI scanner typically incorporates a donut-shaped liquid helium dewar with many kilometers of superconducting wire inside. By passing high currents through these wires, the different magnetic fields necessary to produce an image are generated (e.g. Fig. 5.1). The most characteristic of the magnetic fields is the static measurement field (herein referred to as B_0). In a typical MR-scanner, B_0 is around 1.5 T but the standard is slowly moving towards 3 T. However, there are MR-scanners with B_0 fields over 10 T [4]. The reason for the trend of increasing B_0 is to achieve higher spatial resolution [4].

Ultra-low field MRI (ULF-MRI) with SQUID detection was pioneered by the group of John Clarke in the late 1990s [5, 6]. MRI had been done in low fields (mT) before [7, 8], but now the focus was to decrease the measurement field to microteslas (μT). Since then, similar developments have been made in a number of laboratories [9–12].

There are several advantages of ULF-MRI. Firstly, by going to the very low magnetic fields, metal implants and pacemakers are not a problem, as is the case in conventional MRI. Furthermore, an MR-scanner is loud and the space inside the

F. Öisjöen, *High-T$_c$ SQUIDs for Biomedical Applications: Immunoassays, Magnetoencephalography, and Ultra-Low Field Magnetic Resonance Imaging*, Springer Theses, DOI: 10.1007/978-3-642-31356-1_5, © Springer-Verlag Berlin Heidelberg 2013

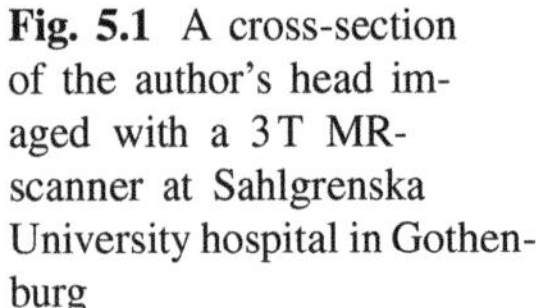

Fig. 5.1 A cross-section of the author's head imaged with a 3 T MR-scanner at Sahlgrenska University hospital in Gothenburg

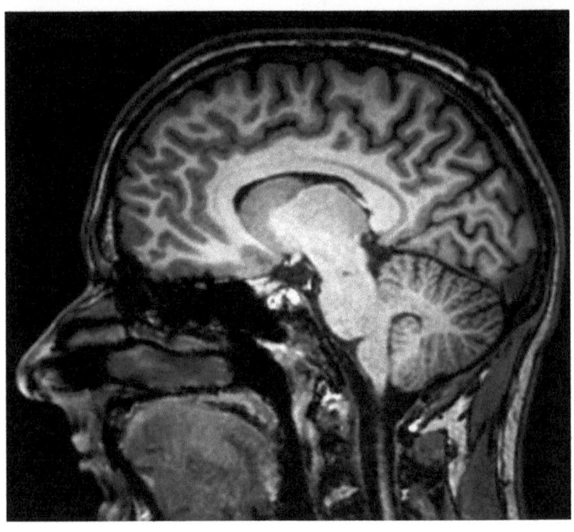

magnet is limited, which may be lead to an uncomfortable experience for the patient. In fact, studies show that more than 20 % of the population cannot undergo an MR due to claustrophobic stress [13]. The cramped space in an MR-scanner is also a problem for obese people. Secondly, the cost of magnets (design and manufacturing) is reduced considerably since the coils can be made with copper wire. This means no liquid helium is needed for cooling the magnets. Furthermore, the magnetic field homogeneity requirements are relaxed at lower fields. Lastly, the contrast in the longitudinal relaxation rate T_1 is enhanced in fields below $\sim$1 mT [14], which has been used for imaging (ex-vivo) of prostate tumors without a contrast agent [15]. This is not possible with conventional MRI. The disadvantage with ULF-MRI is the poor spatial resolution compared with conventional MRI.

For functional brain mapping with MEG (for high temporal resolution) one has to map the MEG/EEG reconstruction onto an anatomical image obtained by MRI. This means two types of apparatus are needed: an MEG system and an MR-scanner, and they are both a huge investment. Furthermore, the subject needs to move between recording brain activity (MEG) and taking the anatomical image (MRI). This leads to inaccuracy in localizing the active areas (neural activity) in the brain. A simple example is that the brain moves a few millimeters between having a subject sitting up (MEG) and laying down (MRI). For brain mapping one wants high spatial and temporal resolution. With the present state-of-the-art MEG and MRI, a combination of the two is impossible due to the high magnetic fields in MRI.

The motivation for our research is driven by the MEGMRI proposal[1]: a hybrid MEG and MRI modality where MRI would be performed in microtesla fields. This would enable co-registration of MEG with ULF-MRI [16] and may increase the accuracy of MEG source reconstruction.

---

[1] http://www.megmri.net

This chapter presents the preliminary results and status of the ULF-MRI development. At the time of writing, images have not yet been obtained but suggestions on how to proceed are discussed. In short, the present achievements are NMR measurements on water samples and one-dimensional imaging with a single gradient.

5.2 Basic Concepts of MRI

The theory and physics behind MRI is well understood and is too extensive to be covered in this thesis. Therefore, only the basic concepts of MRI will be introduced. For a complete description, and as a reference to MRI, see [17].

MRI is based on the interaction between the spin of a nucleus and an external magnetic field, known as nuclear magnetic resonance (NMR). A proton, the nucleus of hydrogen, is the nucleus of interest in MRI because of the abundance of water in the soft tissues of the human body. A spin 1/2 particle (e.g. a proton), placed in an external magnetic field, will have a tendency to align itself parallel or anti-parallel to the magnetic field. In NMR, the total magnetic moment of an ensemble of spins is measured. This means that there has to be a small excess of spins aligned in one direction. The preferred alignment corresponds to the lowest energy state (parallel alignment). This can be shown by considering the difference in Boltzmann distributions of the two states (see [17]). The total equilibrium magnetization of an ensemble of protons with density ρ placed in a magnetic field B at temperature T is

$$M_0 = \frac{\rho \gamma^2 \hbar^2}{4k_B T} B \tag{5.1}$$

where $\gamma/2\pi$ is the gyromagnetic ratio. For protons $\gamma/2\pi = 42.58\,\text{MHz/T}$. The frequency associated with transitions between the two states is given by the Larmor frequency

$$f_L = \frac{\gamma}{2\pi} B. \tag{5.2}$$

In a more classical picture, the spin can be considered a spinning gyroscope. Due to charge of the proton and the spinning motion, it produces its own magnetic field. If the spin is exposed to an external magnetic field perpendicular to the proton magnetic moment (i.e. the axis of the spin motion) it will try to align itself with the magnetic field. However, the proton magnetic moment vector follows a gyroscope-like motion with precession around the axis of the magnetic field before it aligns with the magnetic field. The frequency of the precession is the Larmor frequency (Eq. 5.2).

In a typical NMR/MRI sequence, one rotates the magnetization vector away from the direction of an applied static magnetic field. This is done by applying an excitation pulse with a frequency that matches the Larmor frequency of the spins in the static magnetic field. This is the resonance phenomenon of NMR. In the reference frame of the spins, this field is static and the spins will precess around the direction of the excitation field. With an appropriate duration of the applied excitation one can rotate the spins through an angle $\pi/2$ away from the static magnetic field. This starts the precession (with the Larmor frequency) of the proton spin magnetization about the axis of the static field and can be detected by a receiver coil.

If the proton spins are tipped by a $\pi/2$-pulse into the x-y plane, with the static magnetic field in the z-direction, then the longitudinal magnetization (in the z-direction) will relax back to the z-direction with a characteristic time T_1. This is also called spin-lattice relaxation and the regrowth of the magnetization in the z-direction as a function of time (t) is given by

$$M_z(t) = M_0 \left(1 - \exp\left(-t/T_1\right)\right). \tag{5.3}$$

The transversal magnetization (in the x–y plane) is initially (after a $\pi/2$-pulse) M_0, and by spin dephasing decays with a characteristic time T_2. This relaxation has two components, the first one is due to local differences in the magnetic field caused by spin-spin interactions and is referred to as simply T_2. This is an intrinsic effect. The other one, T_2', is caused by inhomogeneities in the applied magnetic field. The local differences in magnetic fields naturally give a spread in Larmor frequencies over the ensemble of protons and therefore a dephasing effect. The total relaxation caused by dephasing is the combination of the intrinsic (T_2) and extrinsic (T_2'), and is commonly referred to as T_2^*,

$$\frac{1}{T_2^*} = \frac{1}{T_2'} + \frac{1}{T_2}. \tag{5.4}$$

The transversal magnetization decays exponentially after excitation as

$$M_{x,y}(t) = M_0 \exp\left(-t/T_2^*\right). \tag{5.5}$$

Strictly speaking, the magnitude of the initial magnetization in the transversal plane is equal to the longitudinal magnetization just before excitation.

The dephasing caused by field inhomogeneities (T_2') can be reversed by implementing a so called spin-echo sequence but the effect from spin-spin interactions (T_2) is irreversible. The linewidth at full-width-half-maximum (FWHM) Δf of the resonance peak of an NMR spectrum reflects T_2^* through the relationship

$$\Delta f = \frac{1}{\pi T_2^*}. \tag{5.6}$$

It is desirable to minimize the linewidth of the NMR resonance peak in order to achieve high spatial resolution in MRI. This is accomplished by having a highly homogeneous measurement field and is challenging to achieve. A spin-echo sequence enables a direct measurement of T_2.

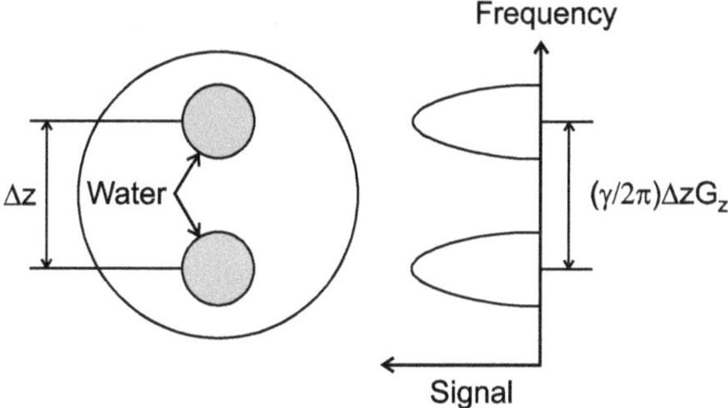

Fig. 5.2 A simple schematic showing the principle of frequency encoding. The spatial separation of the two water columns is directly reflected in frequency space due to the applied gradient G_z

In order to obtain an image, the spins are manipulated in two ways: the frequency and the phase of the precession. Since in this chapter, only frequency-encoding is demonstrated, phase encoding is left out (for a more thorough treatment see [17]). In frequency encoding, a gradient is applied that locally changes the frequency of the spins. From Eq. 5.2 with an added gradient we obtain

$$f(z) = \frac{\gamma}{2\pi}\,(B_0 + zG_z)\,, \tag{5.7}$$

where z is the spatial coordinate and $G_z = dB_z/dz$ is the applied gradient. As an example, we can consider a water phantom with two columns, as shown in Fig. 5.2. The peak frequency of the spins in the two water columns will be separated by $\Delta f = \Delta z G_z \gamma/2\pi$ due to the applied gradient. The widths of the peaks reflect the spatial extension of the water columns.

5.3 NMR/MRI in Ultra Low Fields Using SQUIDs

Imaging in ultra low fields is not conceptually different from conventional MRI. However, one significant difference is the type of detector. The receiver coils in conventional MRI are Faraday type detectors based on induction. This means the measured voltage (V) is directly proportional to the frequency of the measured field, i.e. the magnetization vector precessing with the Larmor frequency, $V \propto dM/dt \sim B_0 f_0$. Furthermore, the Larmor frequency is proportional to the static measurement field (Eq. 5.1). Therefore, the measured voltage scales as the measurement field squared ($V \propto B_0^2$). This is why the current trend in MRI is to increase the static field rather than going to lower fields [4]. For SQUID detected NMR/MRI, the situation is different. The sensitivity of a SQUID sensor is frequency independent and the previous analysis now yields $V \propto B_0$. Furthermore, pre-polarizing the sample in a higher magnetic field B_p increases the SNR but the precession frequency is set by the lower measurement field (B_0). In this case, $V \propto B_p$ and not the lower measurement field

(B_0). With this technique the linewidth (i.e. the spectral resolution) is improved [18]. Pre-polarization is necessary in order to obtain sufficient SNR for ULF-NMR/MRI.

The increased linewidth caused by the inhomogeneities in B_0 is given by $\Delta f' = \gamma/2\pi \, (\Delta B_0/B_0) \, B_0$ where ΔB_0 is the change in field across the sample. This is what causes the extrinsic T_2' decay. As described in [19], the above mentioned relationship between the linewidth and the measurement fields leads to a significant advantage at lower fields, for a given fractional inhomogeneity ($\Delta B_0/B_0$). For example, to achieve a 1 Hz linewidth in a measurement field of 3 T (as the scanner that produced Fig. 5.1) a homogeneity of $\sim$1 ppb is required. In contrast, in a measurement field of 3 mT only $\sim$ 1 ppm is needed for the same linewidth.

5.4 Experimental Methods

This section provides the details of the experimental setup and design. The different parts of this section will describe the coil design, hardware used, and SQUID setup. The experiments and results are further presented in [20].

5.4.1 Experimental Setup

The ULF-NMR/MRI system was setup in an rf-shielded room with a shielding factor of 100 dB at 10 kHz. Data measured by the SQUID was acquired with a data acquisition card (NI-DAQ 6259) that was also used for controlling the pulse sequences for NMR. The output signal from the SQUID electronics output was filtered and amplified (high-pass filter at 300 Hz with a gain of 10). The hardware used for each of the different set of coils is described in the corresponding section. A SQUID magnetometer (MAG3L) was used for the NMR/MRI experiments. The SQUID was placed in a superconducting shield, immersed in liquid nitrogen and further shielded by two layers of mu-metal. A tuned pick-up circuit was used to detect the NMR signal. This circuit consisted of a pick-up copper coil that surrounded the sample (e.g. water) connected to an input copper coil close to the SQUID magnetometer. A capacitor was placed in series with the two coils in order to tune the resonant frequency to the Larmor frequency.

5.4.2 Coil Design

For a full ULF-MRI system a number of coils are required. First of all a static measurement field coil, B_0, is needed to define the Larmor frequency. Three sets of gradient coils for imaging: dB_z/dz provides frequency encoding in the z-direction, and dB_z/dy and dB_z/dx provide phase encoding in the other two directions. Furthermore, cancellation of the Earth's magnetic field in three directions is

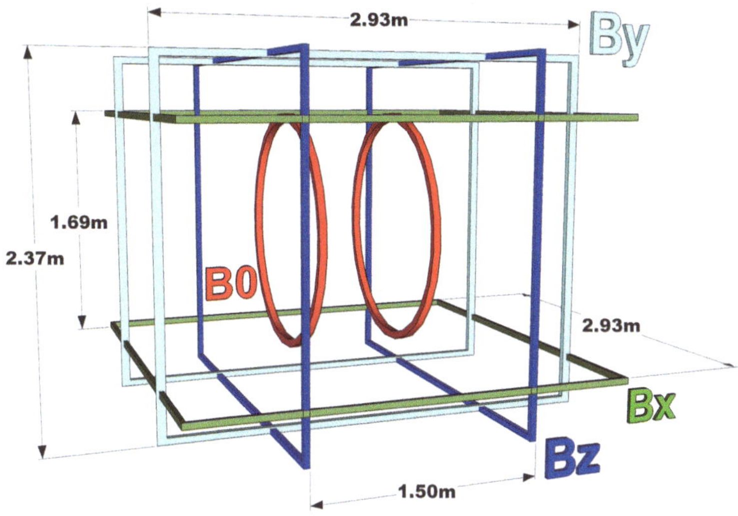

Fig. 5.3 A schematic drawing showing the three axis Helmholtz like cancellation coils (B_x, B_y, B_z) and the 1.6 m diameter measurement (Helmholtz) coil (B_0). For measurements, the sample is placed in the center of the B_0 coils where the field is most homogeneous

necessary (cancellation in the z-direction is not needed if the strength of that component is known). A Helmholtz coil configuration was chosen when applicable due to the homogeneity of the magnetic field produced by such a coil system. A schematic of the coil setup in our system is shown in Fig. 5.3 and includes: B_0, B_x, B_y, B_z. Not shown is a small model system with two gradient coils (dB_z/dz and dB_z/dy) that is placed inside the B_0 coil. This configuration was only designed for imaging of small samples ($\sim$cm). A set of photographs of the ULF-NMR/MRI system is shown in Fig. 5.4. The coil supports were all made in wood and/or plastic in order to avoid any metal parts in the vicinity of the sample volume. The different coils are described separately in the following sections. A summary of the coils can be found in Table 5.1.

Earth's Field Cancellation Coils

The Earth's field cancellation coils (B_z, B_y, B_z) were attached on the inside of the walls of the rf-shielded room in order to maximize the volume of high magnetic field homogeneity. The B_x coils were a square Helmholtz-like pair with dimensions $2.93 \times 2.93 \, \text{m}^2$, separated by 1.69 m. The B_y and B_z coils were also of Helmholtz type but with a rectangular shape due to the geometry of the rf-shielded room. The lengths of the sides were $2.93 \times 2.37 \, \text{m}^2$ and the pairs were separated by 1.50 m. The coils of the Helmholtz configurations had 20 turns each and provided roughly

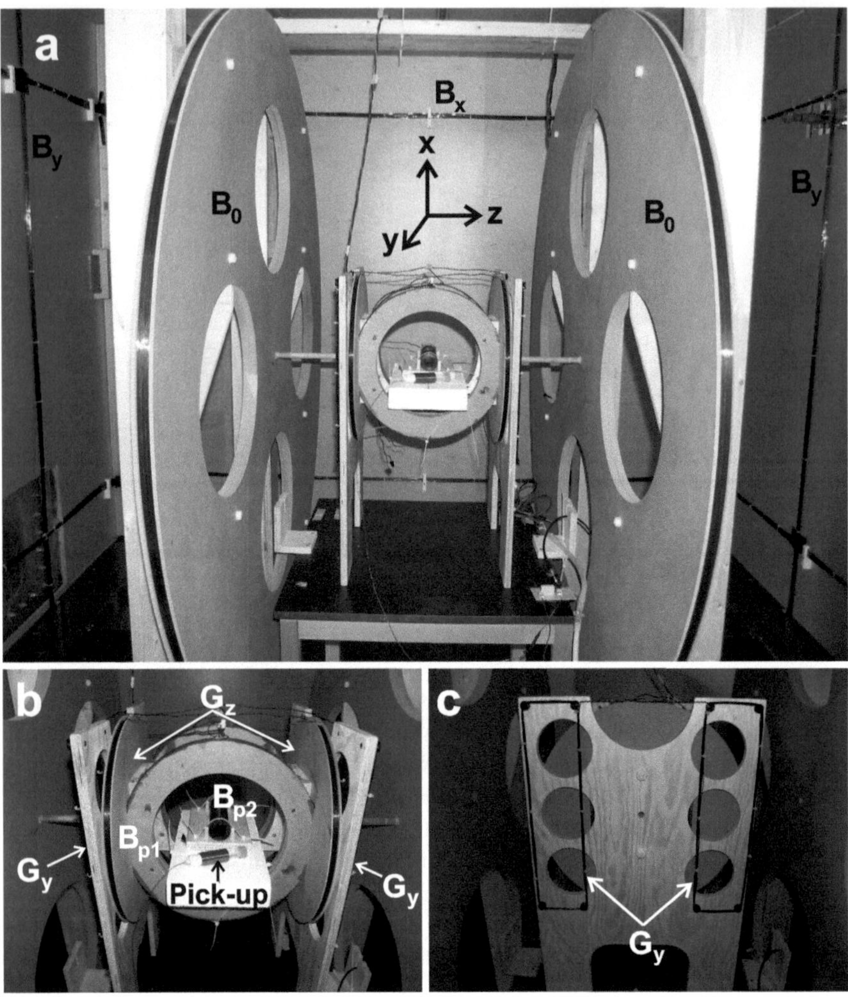

Fig. 5.4 Photographs showing the NMR/MRI setup. **a** The full system with the measurement coils (B_0) and part of two of the pairs of cancellation coils (one of B_x, and one of B_y); **b** A closer look of the model gradient coils where the Maxwell coil (diameter 0.4 m) G_z can be seen. A set of planar coils outside the Maxwell pair constitute G_y. The solenoid B_{p2} used for the NMR experiments is placed in the center, and the indicated pick-up coil that surrounds a water sample fits inside the solenoid. The pick-up coil is directly coupled to an input coil that couples to the SQUID magnetometer (not shown). Also shown is the first pre-polarizing coil B_{p1} (not used). **c** A sideview showing one set of the planar gradient coils with height 0.4 m, width 0.1 m and separated by 0.22 m. There is another identical pair on the opposite side of the setup

10μT/A. The three pairs of cancellations coils were driven by one BK Precision 1,745 A current supply each and the coils were low-pass filtered (-3 dB at 2 Hz).

Table 5.1 A summary of the different coils used for ULF-NMR/MRI

Coil	Coil constant	No. turns, wire gauge	Resistance, inductance	Dimensions (m)	Inhomogeneity
B_x	$10\,\mu\text{T/A}$	20, AWG12	1.6/10	2.93×2.93	7×10^{-5} (70 ppm)
B_y	$10\,\mu\text{T/A}$	20, AWG12	1.5/10	2.93×2.37	2×10^{-4} (0.2 ppt)
B_z	$10\,\mu\text{T/A}$	20, AWG12	1.5/10	2.93×2.37	2×10^{-4} (0.2 ppt)
B_0	$117\,\mu\text{T/A}$	100, AWG14	10/80	1.6	2×10^{-4} (0.2 ppt)
B_{p1}	$1\,\text{mT/A}$	195, AWG14	3.4/16	0.33	–
B_{p2}	$4.5\,\text{mT/A}$	450, 1 mm	1.0/7	12.4×2.5 (cm)	–
G_z	$400\,\mu\text{T/A}$	20, 1 mm	1.1/7	0.4	$7.5\,\mu\text{m}$
G_y	$85\,\mu\text{T/A}$	10, 1 mm	0.9/0.43	$0.4 \times 0.1 \times 0.22$	$10\,\mu\text{m}$

Resistances and inductances are given in Ω and mH, respectively. The homogeneity of the gradient coils (sample volume $5 \times 5 \times 5\,\text{cm}^3$) are calculated from $\Delta B/G$ [21]. The fractional inhomogeneity ($\Delta B/B$) of the static magnetic field coils are calculated in a $10 \times 10 \times 10\,\text{cm}^3$ cube. The dimensions of B_0, B_{p1}, and dB_z/dz specify the diameter. The dimensions of B_{p2} are the length and the inner diameter of the solenoid

The Earth's magnetic field was cancelled inside the measurement volume before the NMR/MRI measurements began. The absolute magnetic field in the center of the setup was measured with a three axis fluxgate magnetometer (Mag639, Barlington Instruments).

Measurement Field and Pre-Polarization Coils

The measurement field (B_0) coil was a circular Helmholtz coil with diameter 1.6 m (and therefore separated by 0.8 m). The two coils of the Helmholtz pair had 100 turns each and produced $117\,\mu\text{T/A}$. Simulations showed that the field inhomogeneity in a $10 \times 10 \times 10\,\text{cm}^3$ cube was 2×10^{-4}. An HP6030 current supply provided the dc current for the B_0 coil. A low-pass filter ($-3\,\text{dB}$ at 1 Hz) was inserted between the power supply and the coil.

Two types of pre-polarizing coils were investigated. The initial idea was to use a Helmholtz-type coil. The homogeneity of the magnetic field in the center would allow us to align a SQUID gradiometer parallel to the field lines, and thus minimize the coupling between the SQUID and the pre-polarizing pulse. The first coil (B_{p1}) was hence a Helmholtz coil with diameter 33 cm. Each of the coils of the Helmholtz pair had 195 turns (AWG 12 Cu wire) and they produced $1\,\text{mT/A}$.

The second pre-polarizing coil (B_{p2}) that was made was a solenoid type coil, 15 cm long with four layers of wire (1 mm, Cu). With a total of 450 turns it could produce $4.5\,\text{mT/A}$. This coil was used to obtain the NMR/MRI results presented in this chapter.

The pre-polarizing pulse was programmed with a Fluke AWG-220 waveform generator and a Techron LVC5050 linear amplifier supplied the current.

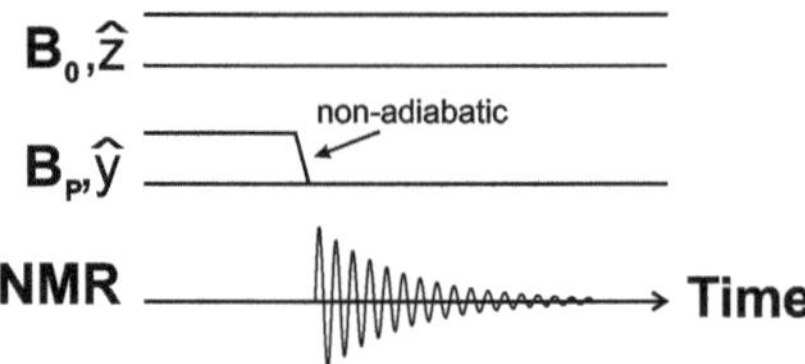

Fig. 5.5 Simple NMR sequence for FID measurements. The static measurement field (B_0) is constantly on in the z-direction. The pre-polarizing pulse (B_p) is applied perpendicular to the measurement field. Non-adiabatic turn-off of B_p causes the spins to precess in B_0

Gradient Coils

The simplest of the gradient coils is the $G_z = dB_z/dz$ gradient. This gradient can be accomplished with a Maxwell coil. This type of coil is similar to a Helmholtz coil, but the two coils are separated by the radius (of the coils) times 1.73 (1.73R) and not the radius itself. The current in the two coils flow in opposite directions (i.e. the coils wound in opposite directions but placed in series) which creates the gradient. The small model Maxwell coil in our model system had diameter 40 cm and were separated by 34.6 cm and produced 400 μT/m with 1 A in the coil. The gradients for the other directions are more complicated. The $G_y = dB_z/dy$ gradient in our model system is produced by a set of four rectangular coils (similar to e.g. [9]). Their geometry was optimized by minimizing the inhomogeneity in a cubic ($5 \times 5 \times 5$ cm^3) sample volume [21] using Matlab.

5.5 Results and Discussion

The results presented were obtained with the solenoid B_{p2} and the copper coil flux transformer. Issues related to the configuration with the SQUID inside B_{p1}, and possible solutions are discussed.

5.5.1 Free Induction Decay and T_1

A simple pulse sequence is to pre-polarize, then switch off the pre-polarizing field non-adiabatically (i.e. faster than the inverse of the Larmor frequency in B_0) and measure the free induction decay (FID) of the spin magnetization, as schematically shown in Fig. 5.5. The average obtained from 20 ml of water after 50 averages is shown in Fig. 5.6. In this measurement, $B_p - 15$ mT was applied for 5 s in each pulse and was switched off non-adiabatically. The resonant peak in Fig. 5.6 with an SNR of 23 appears at 3,472 Hz, close to what was expected from Eq. 5.2 with $B_0 = 81$ μT. By repeating the experiment at different measurement fields (B_0's) we

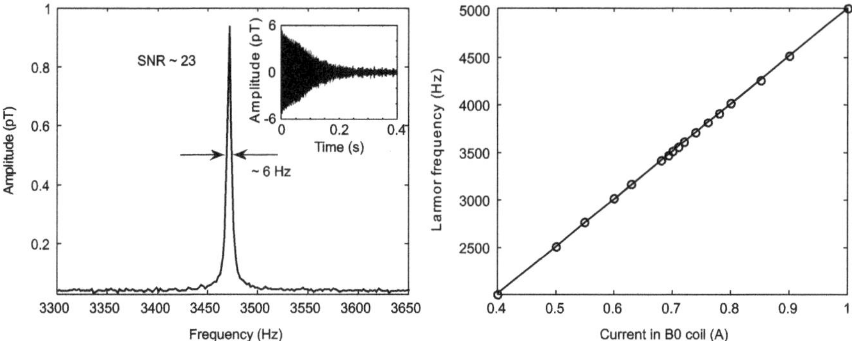

Fig. 5.6 *Left* NMR spectrum obtained from 20 ml of water after 50 averages with $B_p = 15\,\text{mT}$, $B_0 = 81\,\mu\text{T}$. The SNR is 23 and the linewidth 6 Hz. The *inset* shows the time trace of the averaged FID. *Right* The measured Larmor frequency versus the current in the B_0 coil. The linear relationship confirms Eq. 5.2

can verify Eq. 5.2 as a validation. Figure 5.6 shows a plot of the Larmor frequency as measured versus the current in the B_0 coil. The fitted straight line verified the linear relationship of Eq. 5.2 and provided an accurate field/ampere relationship of our measurement coil (B_0).

We estimated the longitudinal relaxation time (T_1) for water by varying the duration of the pre-polarizing time. The initial amplitude of the FID in a certain NMR experiment depends on the time (up to saturation) and amplitude of the pre-polarizing (B_p). The initial amplitude of the magnetization is

$$M_z = M_0 \left(1 - \exp\left(-t_{Bp}/T_1\right)\right). \tag{5.8}$$

where t_{Bp} is the length of the B_p pulse. In order to obtain T_1 for water, we varied the length of the pre-polarizing pulse in the range 0.5–7 s. A plot of the obtained amplitudes versus t_{Bp} is shown in Fig. 5.7. With a fit to Eq. 5.8 we obtained $T_1 \sim 3$ s.

5.5.2 1-D Gradient

We prepared a sample that contained two water volumes of 3 ml each (roughly 1.5 cm long at the edges inside the pick-up coil) separated by 4 cm. This means the center-to-center distance was 5.5 cm. This sample was placed inside B_{p2} and a gradient (dB_z/dy) was applied along the same direction as B_{p2} using the G_y coils. According to Eq. 5.7, we should then be able to resolve two peaks since the two separated volumes of water were individually located in different local magnetic fields.

We performed several measurements with varying amplitude of the applied gradient. In Fig. 5.8 the NMR spectra obtained with gradients up to $6.8\,\mu\text{T/m}$ are shown. The first thing to note in this figure is that even with zero applied gradient we re-

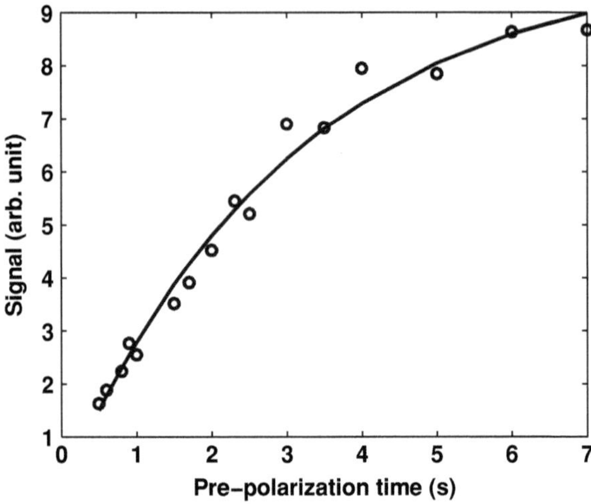

Fig. 5.7 The amplitude of the NMR obtained on 10 ml of water with $B_p = 15$ mT versus different pulse durations (t_{Bp}). The exponential fit (*solid line*) to the data points (*circles*) gives $T_1 \sim 3$ s

solved two peaks. This is an indication that there was an inherent gradient in the system. When we increased the applied gradient to $1.2\,\mu$T/m the peaks almost merged which indicates the setup had a background gradient of roughly $1.2\,\mu$T/m. When the gradient was increased further, the two NMR peaks became more separated, as can be seen in Fig. 5.8. We account for the excess background gradient by subtracting $1.2\,\mu$T/m from all applied gradients. For example, when we applied $G_y = 6\,\mu$T/m the actual gradient of $G_y = 6 - 1.2 = 4.8\,\mu$T/m. Using this gradient we could estimate $\Delta y = \Delta f/(\gamma/2\pi) * G_y \approx 4.5$ cm which reflects the 5.5 cm separation of the centres of the two water volumes. The discrepancy between the 5.5 cm center-to-center separation of the water samples and that estimated with the gradient fields can be explained by the geometry of the samples versus the pick-up coil. The two volumes of water were in fact at the edges of the pick-up coil which means they were placed in the least sensitive part of the coil. It is therefore likely that the effective volume of water that the pick-up coil sensed was much less than the 3 ml (or 1.5 cm extended water sample), thus, the separation as seen by the NMR measurement is smaller.

5.6 Issues and Future Work

During the course of our ULF-MRI development we discovered fluctuations in the magnetic field inside the rf-shielded room that made the apparent measurement field fluctuate in time. A consequence is that the Larmor frequency fluctuated and made averaging difficult. These fluctuations ($\sim$dc drift) was on the order of ± 200 nT.

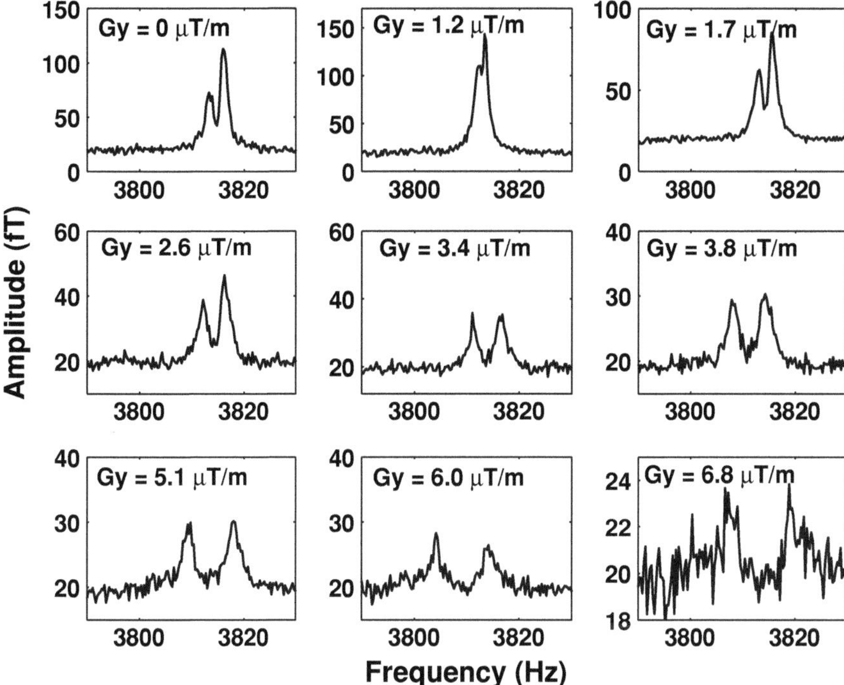

Fig. 5.8 NMR spectra of two water volumes separated by 4cm measured at $B_0 = 81\,\mu$T and varying gradient $G_y = dB_z/dy$. The indicated gradient strength in each figure is the applied gradient. In *the upper left figure*, we see that although the gradient is zero, there are two peaks. This indicates that there was a background gradient present. As the gradient was increased, the NMR peaks first merge ($G_y = 1.2\,\mu$T/m) and then split for higher gradients

The fluctuations were essentially zero at night and there was a quiet window that corresponded with the lack of traffic at that time. Therefore we believe they are the main source of these fluctuations [20]. Figure 5.9 shows 2 single shot NMR peaks where a spread in Larmor frequencies of 6 Hz can be seen. This corresponds to approximately 140 nT according to Eq. 5.2. Measurements at night (at 2 am) did not show any significant drift in Larmor frequency. The fluctuating magnetic field in the measurement environment caused an increased linewidth of the NMR peak when averaging was performed. By manually aligning the individual single shot peaks, or by measuring at night, the linewidth was reduced from 6 to 4 Hz. Although this is an improvement, 4 Hz is still a rather large linewidth.

One possible solution to the issue of the magnetic fluctuations is to use active compensation. Presently, we have the three axis compensation (B_x, B_y, B_z) for the Earth's magnetic field (only dc). These can be extended to actively cancel the fluctuations using a three axis fluxgate magnetometer as a null detector. The second possibility is to measure the fluctuations with a fluxgate magnetometer simultaneously as the NMR experiments and shift the NMR peaks accordingly.

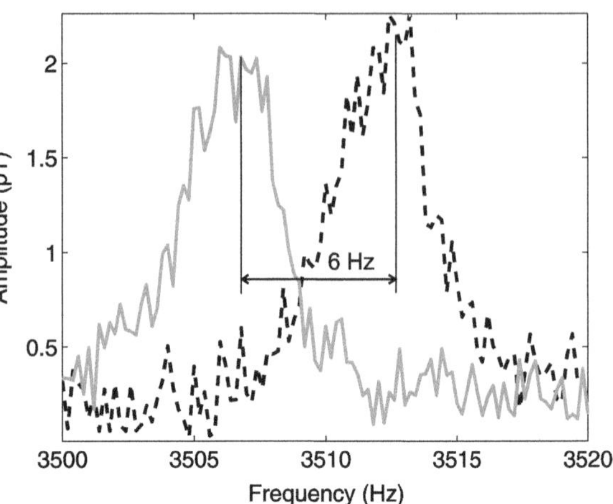

Fig. 5.9 Two single shot NMR signals acquired during the same averaging session. The spread in Larmor frequency causes an increased apparent linewidth when averaging is performed. The difference in Larmor frequencies can be explained by the external fluctuations of roughly $\pm 200\,$nT

A discussed above, during the first attempts to resolve two volumes of water by using a single gradient in the y-direction we discovered a gradient of the magnetic field. As shown in the first figure (upper left) of Fig. 5.8, we were able to resolve the two NMR peaks corresponding to two different water volumes without any applied gradient. This showed that there was an inherent gradient in the system during the measurement. This gradient was another source (apart from the fluctuations) that caused an increased linewidth of the NMR of a single water volume (e.g. Fig. 5.6).

We rotated the two water samples (thus also B_{p2} and the pick-up coil) 90 degrees such that the samples were pre-polarized in the x-direction (see Fig. 5.4). We performed an NMR (30 averages, and $B_p = 15\,$mT) on The same sample as in the gradient experiment (2 volumes with 3 ml of water each) in this direction and implemented the shift of NMR peaks according to fluctuations in the ambient field tracked by the fluxgate magnetometer. In Fig. 5.10, the direct average (solid line) of the NMR is shown and the average of the aligned NMR peaks (dashed line). Although the sample contained two volumes of water there is only one NMR peak. This showed that there is no (or only a weak) inherent gradient in the x-direction. Furthermore, the linewidth decreased from 5 to 1.5 Hz by aligning the peaks before averaging. A 1.5 Hz linewidth corresponds to $T_2^* \sim 210\,$ms.

The long term goal is to place the SQUID in direct proximity of the sample, i.e. eliminate the copper flux transformer and fully take advantage of the small sensor-to-sample separation enabled by high-T_c technology. This means the SQUID will inevitably have to be placed inside the pre-polarizing field. This was the configuration in the initial setup and lead to limitations on the strength of the pre-polarizing field. The main problem was the coupling between the pre-polarizing field and the SQUID that caused significant flux trapping in the superconducting film of the SQUID. A possible solution to this problem is to heat the superconducting film to above T_c during, or right after the pre-polarizing pulse and then rapidly cool it down before acquisition. The main factor for a real implementation is to heat the superconducting

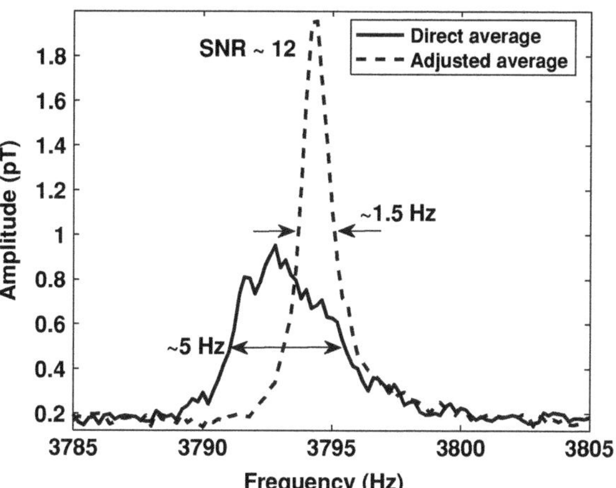

Fig. 5.10 Two NMR spectra of 6 ml of water (2 volumes, 3 ml each, separated by 4 cm) averaged 30 times with $B_p = 15$ mT. *The solid line* NMR was obtained by making a simple direct average. *The dashed line* represents the NMR averaged after aligning the peaks of each individual NMR according to background measurements of the external fluctuations. The linewidth after alignment was 1.5 Hz and the SNR was 12 compared to 6 Hz and SNR 6 without alignment

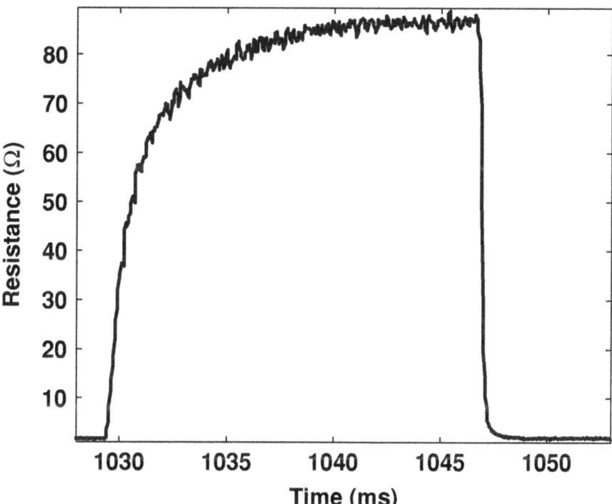

Fig. 5.11 The resistance of the SQUID as a function of time as a laser pulse hits the surface of it. The cooling time, starting from roughly 1,047 ms is less than 5 ms. The resistance of the SQUID increased by a factor of 40 in 10 ms indicating fast heating to above T_c

film to above T_c without injecting more energy into the system than necessary. This minimizes the cooling time since the cooling power is set by the cryostat.

Initial attempts using resistive heaters [22] heated more than the superconducting film, which made the cooling time long (100 ms) due to the excessive amount of energy needed. This led us to try laser heating [23] with the motivation that with the correct wavelength only the superconducting film (YBCO, absorption peak at 1,800 nm [24]) would absorb the energy. With a simple setup we continuously measured the resistance of the SQUID as we shone a laser beam on it. We placed a simple chopper between the SQUID and the laser output. The SQUID was placed in a cryostat (described in e.g. Sect. 4.4.1) where it was separated from room temperature by a sapphire window that was roughly transparent to infra-red radiation. A time trace of one such recording with a laser power of 100 mW and wavelength 1,300 nm is shown in Fig. 5.11. The total heating and cooling time is roughly 20 ms. More importantly, the cooling time is less than 5 ms, sufficient for NMR/MRI implementation. Although this experiment was not optimized, it showed the proof-of-principle.

In the more immediate future, we will implement the possibility to do excitation in our system. This will be accomplished using a Helmholtz type excitation coil. The excitation coil will produce $\pi/2$ and π pulses, which reflects the angle of rotation of the spins [17]. This enables for example spin echo sequences with which one can measure the intrinsic T_2.

References

1. P. Lauterbur, Image formation by induced local interactions: examples employing nuclear magnetic resonance. Nature **242**, 190–191 (1973)
2. F. Bloch, W.W. Hanson, M. Packard, The nuclear induction experiment. Phys. Rev. **70**, 474–485 (1946)
3. E.M. Purcell, H.C. Torrey, R.V. Pound, Resonance absorption by nuclear magnetic moments in a solid. Phys. Rev. **69**, 37–38 (1946)
4. H. Wada, M. Sekino, H. Ohsaki, T. Hisatsune, H. Ikehira, T. Kiyoshi, Prospect of high-field MRI. IEEE Trans. Appl. Supercond. **20**, 115–122 (2010)
5. M.P. Augustine, A. Wong-Foy, J.L. Yarger, M. Tomaselli, A. Pines, D.M. Ton That, J. Clarke, Low field magnetic resonance images of polarized noble gases obtained with a dc superconducting quantum interference device. Appl. Phys. Lett. **72**, 115–122 (1998)
6. D.M. TonThat, M. Ziegeweid, Y.-Q. Song, E.J. Munson, S. Appelt, A. Pines, J. Clarke, SQUID detected NMR of laser-polarized xenon at 4.2 K and at frequencies down to 200 Hz. Chem. Phys. Lett. **272**, 245–249 (1997)
7. Y.S. Greenberg, Applications of SQUIDs to NMR. Rev. Mod. Phys. **1**, 175–222 (1998)
8. H.C. Seton, J.M.S. Hutchison, D.M. Bussell, A 4.2 K receiver coil and SQUID amplifier used to improve the SNR of low-field magnetic resonance images of the human arm. Meas. Sci. Technol. **8**, 198–207 (1997)
9. V.S. Zotev, A.N. Matlachov, P.L. Volegov, A.V. Urbaitis, M.A. Espy, R.H. Kraus, SQUID-based instrumentation for ultralow-field MRI. Supercond. Sci. Technol. **20**, 367–373 (2007)
10. M. Burghoff, H.H. Albrecht, S. Hartwig, I. Hilschenz, R. Körber, T.S. Thömmes, H.J. Scheer, J. Voigt, L. Trahms, SQUID system for MEG and low field magnetic resonance imaging. Metrol. Meas. Syst. **16**, 371–375 (2009)
11. H. Dong, Y. Zhang, H. Krause, X. Xie, A. Offenhäusser, Low field MRI detection with tuned HTS SQUID magnetometer. IEEE Trans. Appl. Supercond. **21**, 509–513 (2011)
12. S. Liao, K. Huang, H. Yang, C. Yen, M.J. Chen, H. Chen, H. Horng, S.Y. Yang, Characterization of tumors using high-t_c superconducting quantum interference device detected nuclear magnetic resonance and imaging. Appl. Phys. Lett. **97**, 263701 (2010)
13. H.K. McIsaac, D.S. Thordarson, R. Shafran, S. Rachman, G. Poole, Claustrophobia and the magnetic resonance imaging procedure. J. Behav. Med. **21**, 255–268 (1998)
14. S.K. Lee, M. Moßle, W. Myers, N. Kelso, A.H. Trabesinger, A. Pines, J. Clarke, SQUID-detected MRI at 132 μT with T_1-weighted contrast established at 10 μT-300 mT. Magn. Reson. Med **53**, 9–14 (2004)
15. S.E. Busch, *Ultra-Low Field MRI of Prostate Cancer Using SQUID Detection*. Ph.D. Thesis, University of California at Berkeley, 2011
16. P.E. Magnelind, J.J. Gomez, A.N. Matlashov, T. Owens, J.H. Sandin, P.L. Volegov, M.A. Espy, Co-registration of interleaved meg and ulf mri using a 7 channel low-T_c system. IEEE. Trans. Appl. Supercond. **21**, 456–460 (2011)
17. E.M. Haache, R.W. Brown, M.R. Thompson, R. Venkatesan, *Magnetic Resonance Imaging: Physical Principles and Sequence Design* (Wiley, New York, 1999)

18. R. McDermott, A.H. Trasbinger, M. Mück, E.L. Hahn, A. Pines, J. Clarke, Liquid-state NMR and scalar couplings in microtesla magnetic fields. Science **295**, 2247–2249 (2002)
19. J. Clarke, M. Hatridge, M. Möble, SQUID-detected magnetic resonance imaging in microtesla fields. Annu. Rev. Biomed. Eng. **9**, 389–413 (2007)
20. M. Jönsson, *Ultra-Low Field Nuclear Magnetic Resonance Using High-t_c SQUIDs*. Master's Thesis, Chalmers University of Technology, 2011
21. W.R. Myers, *Potential Applications of Microtesla Magnetic Resonance Imaging Detected Using a Superconducting Quantum Interference Device*. Ph.D. Thesis, University of California at Berkeley, 2006
22. H.A. Tash, *Fast De-Trapping of Magnetic Flux from High Temperature Superconducting Devices*. Master's Thesis, Chalmers University of Technology, 2010
23. C.E. Cunningham, T.R. DeYoung, T.A. Bouma, Laser light heating for low-noise temperature control in SQUID applications. Phys. B **284**, 2111–2112 (2000)
24. H.L. Dewing, E.K.H. Salje, The effect of the superconducting phase transition on the near-infrared absorption of $YBa_2Cu_3O_{7-\delta}$. Supercond. Sci. Technol. **5**, 50–53 (1992)

Chapter 6
Conclusions

High-T_c (YBCO) dc SQUID sensors were developed and employed in magnetic immunoassays, magnetoencephalography and ultra-low field nuclear magnetic resonance. The noise of the sensors is an important factor since it affects the SNR of any measurement. This leads to e.g. faster MNP measurements (for immunoassays), better localization of MEG sources, and higher resolution in ULF-MRI.

The planar gradiometers for immunoassays showed excellent performance in the unshielded measurement environment. The lowest intrinsic flux noise achieved was $4.6\,\mu\Phi_0/\sqrt{Hz}$ (at 10 Hz) for the SQUID (GRAD1) with a critical current of 125 μA and maximum voltage modulation of 41 μV at 77 K. Several SQUID magnetometers showed excellent performance at the low frequencies where most MEG sources are present. The best magnetic field equivalent noise was 25 fT/$\sqrt{Hz}$ above 40 Hz and 40 fT/$\sqrt{Hz}$ at 10 Hz measured in the MEG setup. Furthermore, a $1/f$-knee at 0.5 Hz was achieved with a high-T_c SQUID gradiometer (GRAD6), a promising result for high-T_c sensor technology.

Biomolecule sensitivity was demonstrated with a system based on the developed high-T_c SQUID gradiometers and magnetic nanoparticles. Binding of biomolecules alters the magnetic relaxation properties of the magnetic nanoparticles that can be detected with the fabricated YBCO dc SQUID gradiometers. The magnetic relaxation dynamics was measured in both the time and frequency domain. In a cluster type assay a sensitivity of 18 ng/ml (roughly 4×10^{10} PSA molecules) was extrapolated and with a one-step type assay $\sim$10 μg/ml was demonstrated. Moreover, MNP content sensitivity of the present setup is 1.5 ng and can be used to extrapolate an ultimate sensitivity to analyte of 4×10^6 corresponding to 3.5 pM or 500 pg/ml with the one-step approach. This can be compared with [1–3]. In these studies, Eberbeck et al. [1] could quantify 2 nM of streptavidin using an MRX approach, Enpuku et al. [3] demonstrated a detection limit of 0.1 pg of IL8 (corresponding to 50 fM) using a sandwich-type assay, and Grossman et al. [2] detected bacteria with a sensitivity corresponding to roughly 50 fM.

The high-T_c SQUID magnetometers were employed as sensors in a two-channel MEG setup that we developed. The well known occipital α-rhythm was recorded

F. Öisjöen, *High-T_c SQUIDs for Biomedical Applications: Immunoassays, Magnetoencephalography, and Ultra-Low Field Magnetic Resonance Imaging*, Springer Theses, DOI: 10.1007/978-3-642-31356-1_6, © Springer-Verlag Berlin Heidelberg 2013

and the characteristic attenuation of the amplitude of the α-rhythm was detected and resolved in time. The amplitude of the recorded occipital α-rhythm was $800\,\mathrm{fT}/\sqrt{\mathrm{Hz}}$. Furthermore, with the two-channel MEG setup we successfully recorded both the occipital α-rhythm and the μ-rhythm in the motor cortex and detect the modulation of both in one and the same recording.

More interestingly, high amplitude θ-band activity was detected in the occipital part of the brain. Such activity is not normally associated with this part of the brain, as recorded with conventional methods. The close proximity of the sensor to the scalp, and therefore to the source inside the brain, may be responsible for this discovery.

Lastly, the first steps towards ULF-MRI were taken. NMR spectra of 20 ml of water were obtained in a magnetic field of $81\,\mu\mathrm{T}$ with an SNR of 23 and linewidth of 6 Hz. Using this system, the longitudinal relaxation time of water was estimated to $\sim$3 s. Furthermore, by aligning the individual resonant peaks of single shot NMRs a linewidth of 1.5 Hz (corresponding to $T_2^* \sim 210$ ms) was obtained with an SNR of 12 on 6 ml of water. Alignment was necessary due to external fluctuations in the magnetic field. Lastly, by applying a gradient $G_y = dB_z/dy$, we could resolve two 3 ml samples of water separated by 4 cm in a one-dimensional imaging experiment.

6.1 Outlook and Discussion

Ever since their discovery, high-T_c SQUIDs have struggled to find their way into a truly successful application on the market. Although the competition is strong, this thesis is aimed to demonstrate how high-T_c SQUIDs can certainly challenge their low-T_c counterparts. The reader is urged to keep in mind that low-T_c SQUIDs have had some 40 years to be developed whereas high-T_c SQUIDs have only been around for roughly half that time.

The work concerning MIA in this thesis was part of a EU project (Biodiagnostics) that finished in 2009. The focus of our activities have since then moved to the field of MEG and ULF-MRI within another EU program (MEGMRI). Nevertheless, activities within the MIA field have continued and recently a high-T_c SQUID and MNP-based assay for detection of markers for Alzheimer's disease in human blood plasma was demonstrated [4]. This was an important demonstration since the most commonly used technique (ELISA) fails at that task.

An immunoassay is not only characterized by the sensitivity to a certain analyte. Other important factors are preparation time and speed of analysis, available sample volume, specificity, and dynamic range. All these factors make it difficult to compare assays and different measurement techniques directly. For example, although the sensitivity of our system was not better than those of other methods (e.g. [2–5]), the one-step assay was simple and fast. Furthermore, it was optimized for small sample volumes in order to increase the relative sensitivity. A highly sensitive assay, in terms of concentration, may fail if the available sample volume is too small. Moreover, using small droplet sample volumes ($\sim\mu$l) together with a microdroplet handling system based on EWOD enables high through-put capabilities. While speed and simplicity come at the cost of extreme sensitivity, such a trade off is often acceptable when rapid screening of many samples against multiple pathogens is needed. Such capability is critical in e.g. cases of potential epidemic of an identified virus of bacterial infection.

Systems employing high-T_c SQUIDs geared towards bio-molecule detection have now reached the market, e.g. MagQU Co. Ltd., XacPro-S101 [6]. This system is marketed as a hyper-sensitive tool for bio-assays based on a magneto-reduction method [7]. Moreover MNPs were proposed [8] and have successfully been used for detection of sentinel lymph nodes with a high-T_c SQUID as the detector and the method is now undergoing clinical trials [9] (trials using SentiMag®, Endomagnetics, UK [10]). This is an important technique since it can be used to locate the first lymph node to which cancer has spread. This lymph node can be surgically removed in order to help prevent further spreading of the cancer.

Naturally, the next step for high-T_c SQUID MEG is to construct a true multi-channel system. The present two-channel system is rather bulky and a new cooling system would be needed that can incorporate multiple SQUIDs in one dewar. The design of such a system could be based on what is presently used for conventional MEG but with the advantage of having the SQUID sensors closer to room-temperature. In order to fully exploit this advantage while keeping the simplicity of a helmet shaped dewar the next system would probably focus on a smaller area of the scalp rather than full head coverage. Before designing such a system, one has to optimize the geometrical layout of the sensors. This involves analyzing the information content available to a given configuration of sensors [11] and will also give more insight into the possible advantage of high-T_c sensors for MEG.

The ultimate goal of our MEG effort is to fully exploit the high operating temperature of high-T_c SQUIDs. Instead of using a rigid helmet shaped dewar like those in state-of-the-art MEG systems, one could construct a flexible array of SQUIDs, mounted in other types of cooling systems. This would enable the SQUIDs to come close to the surface of arbitrary head shapes, which is not possible with the rigid MEG dewars used today. Furthermore, the sensors would be able to sit flat on the surface of the head of a subject. This type of MEG system would also be able to, among other things, eliminate the problem of the head moving with respect to the SQUID sensor array.

Although the noise of our sensors is sufficiently low for basic MEG recordings, more work can improve their performance. Present efforts to develop a reliable fabrication technology for YBCO flux transformers are promising and an improvement of magnetic field noise by a factor of 5 is likely. A flexible array of such high-T_c SQUID sensors could compete with the present state-of-the-art MEG systems.

The ULF-NMR/MRI system is at an early stage of development and some of the required work is discussed in Sect. 5.6 and is related to fluctuations in the ambient magnetic field and rapid de-trapping of magnetic flux from the superconducting film of the SQUID sensors. If we look beyond the immediate future, a combination of MEG and MRI using high-T_c SQUID sensors is a long term goal. This is where de-trapping of flux becomes important since in order to have the SQUID sensors close to the head of a subject, and also pre-polarize for ULF-MRI, the SQUIDs will be exposed to the pre-polarizing magnetic field. This can be solved by making a superconducting flux transformer like those employed in low-T_c SQUID systems [12]. Presently, similar technology for high-T_c is not available. This makes taking full advantage of the small sensor-to-sample separation possible with high-T_c SQUIDs a major challenge. Fast heating and cooling for de-trapping of flux was proposed to be done via laser heating in Sect. 5.6 but has not yet been implemented. High-T_c SQUIDs have been applied to ULF-NMR/MRI by other groups [13–17] but not in a configuration where the pre-polarizing coil is placed arbitrarily with respect to the SQUID.

Another problem is the inductive response of the shielding (an issue that was addressed during the three months I spent working for John Clarke at UC Berkeley) when pre-polarizing with high magnetic fields. The induced currents in the rf-shielding generates a magnetic field that may not only affect the SQUID sensor but also the spins in the sample to be imaged. This issue is treated in [18] and it was shown that by cutting the rf-shielding into smaller pieces, and thereby limiting the path for the induced currents, the situation was improved considerably. If one wants to incorporate MEG into the system, then mu-metal shielding is needed. The risk of magnetizing the mu-metal adds to the complexity and other solutions may be more appropriate [18, 19].

A suggested design for a hybrid MEGMRI high-T_c system can have a nitrogen cooled pre-polarizing coil that surrounds the head of a subject. The SQUIDs could in principle be placed in the same dewar as the pre-polarizing coil but then the small sensor-to-head advantage of high-T_c is likely lost. Instead, the SQUID sensors could be placed in a separate cooling system that enables individual alignment and placement of the SQUID sensors [20, 21]. In proximity to every SQUID sensor one can mount an optical fiber for guiding a laser pulse for fast de-trapping of flux after the pre-polarizing pulse of the ULF-MRI.

Afterword

Hopefully, some of the areas presented in this thesis will spark new development in high-T_c technology. Since our efforts on MIA, systems based on similar methods have reached the market. Our MEG effort is at an early stage but may have market potential with further development and could have a major impact on high-T_c SQUID development. The ULF-NMR/MRI results presented were all obtained in the last month of writing this thesis, hence the preliminary status. While the implementations of the applications presented in this thesis are not exhaustively developed, I hope the advancements made throughout the course of this work can shed new light on the capabilities of high-T_c SQUIDs in the biomedical field.

References

1. D. Eberbeck, C. Bergemann, S. Hartwig, U. Steinhoff, L. Trahms, Quantification of specific bindings of biomolecules by magnetorelaxometry. J. Nanobiotechnol. **6**(4), 13 (2008)
2. H.L. Grossman, W.R. Myers, V.J. Vreeland, R. Bruehl, M.D. Alper, C.R. Bertozzi, J. Clarke, Detection of bacteria in suspension by using a superconducting quantum interference device. Proc. Natl. Acad. Sci. USA **101**(1), 129–134 (2004)
3. K. Enpuku, A. Ohba, K. Inoue, T.Q. Yang, Application of HTS SQUIDs to biological immunoassays. Phys. C **412–414**, 1473–1479 (2004)
4. C. Yang, S. Yang, J. Chieh, H. Horng, C. Hong, H. Yang, K.H. Chen, B.Y. Shih, T. Chen, M. Chiu, Biofunctionalized magnetic nanoparticles for specifically detecting biomarkers of Alzheimer's disease in vitro. ACS Chem. Neurosci. **2**, 500–505 (2011)
5. M. Strömberg, J. Göransson, K. Gunnarsson, M. Nilsson, P. Svedlindh, M. Strømme, Sensitive molecular diagnastics using volume-amplified magnetic nanobeads. Nanoletters **8**(3), 816–821 (2008)
6. http://www.magqu.com/pdt/s101/index.php (Novemb 2011)
7. S.Y. Yang, J.J. Chieh, W.C. Wang, C.Y. Yu, C.B. Lan, J.H. Chen, H.E. Horng, C.-Y. Hong, H.C. Yang, W. Huang, Ultra-highly sensitive and wash-free bio-detection of H5N1 virus by immunomagnetic reduction assays. J. Virol. Methods **153**, 250–252 (2008)
8. S. Tanaka, A. Hirata, Y. Saito, T. Mizoguchi, Y. Tamaki, I. Sakita, M. Monden, Application of high Tc SQUID magnetometer for sentinel-lymph node biopsy. IEEE Trans. Appl. Supercond. **11**, 665–668 (2001)
9. U.A. Gunasekera, Q.A. Pankhurst, M. Douek, Imaging applications of nanotechnology in cancer. Targ. Oncol. **4**, 169–181 (2009)
10. http://www.endomagnetics.com/sentimag/ (Novemb 2011)
11. P.K. Kemppainen, R.J. Ilmoniemi, Channel capacity of multichannel magnetometers, in S.J. Williamson et al. (eds.) *Advances in Biomagnetism* (Plenum Press, New York, 1989), pp. 635–638
12. J. Clarke, M. Hatridge, M. Möble, SQUID-detected magnetic resonance imaging in microtesla fields. Annu. Rev. Biomed. Eng. **9**, 389–413 (2007)
13. H. Dong, Y. Zhang, H. Krause, X. Xie, A. Offenhäusser, Low field MRI detection with tuned HTS SQUID magnetometer. IEEE Trans. Appl. Supercond. **21**, 509–513 (2011)
14. S. Liao, K. Huang, H. Yang, C. Yen, M.J. Chen, H. Chen, H. Horng, S.Y. Yang, Characterization of tumors using high-t_c superconducting quantum interference device detected nuclear magnetic resonance and imaging. Appl. Phys. Lett. **97**, 263701 (2010)
15. K. Schlenga, R. McDermott, J. Clarke, R.E. de Souza, A. Wong-Foy, A. Pines, Low-field magnetic resonance imaging with a high-T_c dc superconducting quantum interference device. Appl. Phys. Lett. **75**, 3695–3697 (1999)
16. O. Snigirev, M. Hayashi, S. Fukumoto, Y. Hatsukade, Y. Katsu, S. Tanaka, Development of ultra low field nuclear magnetic resonance imaging system using HTS rf SQUID. J. Supercond. Nov. Magn. **24**, 1033–1036 (2011)
17. L. Qui, Y. Zhang, H.-J. Krause, A. Braginski, M. Burghoff, L. Trahms, Nuclear magnetic resonance in the earth's magnetic field using a nitrogen-cooled superconducting quantum interference device. Appl. Phys. Lett. **91**, 072505 (2007)
18. K. Zevenhoven, *Solving Transient Problems in Ultra-Low-Field MRI.* Master's Thesis, Aalto University, 2011
19. J.O. Nieminen, P.T. Vesanen, K.C.J. Zevenhoven, J. Dabek, J. Hassel, J. Luomahaara, J.S. Penttilä, R.J. Ilmoniemi, Avoiding eddy-current problems in ultra-low-field mri with self-shielded polarizing coils. J. Mag. Res. **212**, 154–160 (2011)
20. N. Khare, P. Chaudhari, Operation of bicrystal junction high-T_c direct current-SQUID in a portable microcooler. Appl. Phys. Lett. **65**, 2353–2355 (1994)
21. P.P.P.M. Lerou, H.J.M. ter Brake, J.F. Burger, H.J. Holland, H. Rogalla, Characterization of micromachined cryogenic coolers. J. Micromech. Microeng. **17**, 1956–1960 (2007)